Food Sovereignty

A fundamentally contested concept, food sovereignty (FS) has – as a political project and campaign, an alternative, a social movement and an analytical framework – barged into global discourses, both political and academic, over the past two decades. This collection identifies a number of key questions regarding FS. What does (re)localisation mean? How does the notion of FS connect with similar and/ or overlapping ideas historically? How does it address questions of both market and non-market forces in a dominantly capitalist world? How does FS deal with such differentiating social contradictions? How does the movement deal with larger issues of nation-state, where a largely urbanised world of non–food producing consumers harbours interests distinct from those of farmers? How does FS address the current trends of crop booms, as well as other alternatives that do not sit comfortably within the basic tenets of FS, such as corporate-captured fair trade? How does FS grapple with the land question and move beyond the narrow 'rural/agricultural' framework? Such questions call for a new era of research into FS, a movement and theme that in recent years has inspired and mobilised tens of thousands of activists and academics around the world: young and old, men and women, and rural and urban. This book was originally published as a special issue of *Third World Quarterly*.

Eric Holt-Giménez has been the Executive Director of Food First since 2006. He is the editor of many academic, magazine and news articles, and lectures internationally on hunger and food system transformation.

Alberto Alonso-Fradejas is a PhD candidate at the International Institute of Social Studies (ISS) in the Netherlands, a researcher at the Transnational Institute (TNI) and a Fellow of the Guatemalan Institute of Agrarian and Rural Studies (IDEAR).

Todd Holmes is a postdoctoral fellow with the Bill Lane Center for the American West at Stanford University. He earned his PhD in history from Yale University, where for four years he served as the Program Coordinator for the Yale Agrarian Studies Program.

Martha Jane Robbins is a PhD researcher at ISS in the Netherlands and holds a Joseph-Armand Bombardier Canada Graduate Scholarship.

ThirdWorlds

Edited by
Shahid Qadir, *University of London, UK*

ThirdWorlds will focus on the political economy, development and cultures of those parts of the world that have experienced the most political, social, and economic upheaval, and which have faced the greatest challenges of the postcolonial world under globalisation: poverty, displacement and diaspora, environmental degradation, human and civil rights abuses, war, hunger, and disease.

ThirdWorlds serves as a signifier of oppositional emerging economies and cultures ranging from Africa, Asia, Latin America, Middle East, and even those 'Souths' within a larger perceived North, such as the U.S. South and Mediterranean Europe. The study of these otherwise disparate and discontinuous areas, known collectively as the Global South, demonstrates that as globalisation pervades the planet, the south, as a synonym for subalterity, also transcends geographical and ideological frontier.

For a complete list of titles in this series, please visit https://www.routledge.com/series/TWQ

Recent titles in the series include:

Negotiating Well-being in Central Asia
Edited by David W. Montgomery

New Actors and Alliances in Development
Edited by Lisa Ann Richey and Stefano Ponte

Emerging Powers and the UN
What Kind of Development Partnership?
Edited by Thomas G. Weiss and Adriana Erthal Abdenur

Corruption in the Aftermath of War
Edited by Jonas Lindberg and Camilla Orjuela

Everyday Energy Politics in Central Asia and the Caucasus
Citizens' Needs, Entitlements and Struggles for Access
Edited by David Gullette and Jeanne Féaux de la Croix

The UN and the Global South, 1945 and 2015
Edited by Thomas G. Weiss and Pallavi Roy

The Green Economy in the Global South
Edited by Stefano Ponte and Daniel Brockington

Food Sovereignty
Convergence and Contradictions, Condition and Challenges
Edited by Eric Holt-Giménez, Alberto Alonso-Fradejas, Todd Holmes and Martha Jane Robbins

The International Politics of Ebola
Edited by Anne Roemer-Mahler and Simon Rushton

Rising Powers and South-South Cooperation
Edited by Kevin Gray and Barry K. Gills

The 'Local Turn' in Peacebuilding
The Liberal Peace Challenged
Edited by Joakim Öjendal, Isabell Schierenbeck and Caroline Hughes

China's Contingencies and Globalization
Edited by Changgang Guo, Liu Debin and Jan Nederveen Pieterse

The Power of Human Rights/The Human Rights of Power
Edited by Louiza Odysseos and Anna Selmeczi

Class Dynamics of Development
Edited by Jonathan Pattenden, Liam Campling, Satoshi Miyamura and Benjamin Selwyn

Food Sovereignty

Convergence and Contradictions, Condition and Challenges

Edited by
**Eric Holt-Giménez, Alberto Alonso-Fradejas,
Todd Holmes and Martha Jane Robbins**

Routledge
Taylor & Francis Group

LONDON AND NEW YORK

First published 2017 by Routledge

2 Park Square, Milton Park, Abingdon, Oxfordshire OX14 4RN

52 Vanderbilt Avenue, New York, NY 10017

Routledge is an imprint of the Taylor & Francis Group, an informa business

First issued in paperback 2018

British Library Cataloguing in Publication Data
A catalogue record for this book is available from the British Library

ISBN 13: 978-0-415-78634-8 (hbk)
ISBN 13: 978-0-367-11038-3 (pbk)

Typeset in TimesNewRomanPS
by diacriTech, Chennai

Publisher's Note
The publisher accepts responsibility for any inconsistencies that may have arisen during the conversion of this book from journal articles to book chapters, namely the possible inclusion of journal terminology.

Disclaimer
Every effort has been made to contact copyright holders for their permission to reprint material in this book. The publishers would be grateful to hear from any copyright holder who is not here acknowledged and will undertake to rectify any errors or omissions in future editions of this book.

Contents

CONTENTS

Citation Information

The chapters in this book were originally published in *Third World Quarterly*, volume 36, issue 3 (April 2015). When citing this material, please use the original page numbering for each article, as follows:

Chapter 1
Food sovereignty: convergence and contradictions, conditions and challenges
Alberto Alonso-Fradejas, Saturnino M. Borras Jr, Todd Holmes, Eric Holt-Giménez and Martha Jane Robbins
Third World Quarterly, volume 36, issue 3 (April 2015) pp. 431–448

Chapter 2
Exploring the 'localisation' dimension of food sovereignty
Martha Jane Robbins
Third World Quarterly, volume 36, issue 3 (April 2015) pp. 449–468

Chapter 3
Food sovereignty, food security and fair trade: the case of an influential Nicaraguan smallholder cooperative
Christopher M. Bacon
Third World Quarterly, volume 36, issue 3 (April 2015) pp. 469–488

Chapter 4
Food sovereignty and the quinoa boom: challenges to sustainable re-peasantisation in the southern Altiplano of Bolivia
Tanya M. Kerssen
Third World Quarterly, volume 36, issue 3 (April 2015) pp. 489–507

Chapter 5
Food sovereignty as praxis: rethinking the food question in Uganda
Giuliano Martiniello
Third World Quarterly, volume 36, issue 3 (April 2015) pp. 508–525

For any permission-related enquiries please visit:
www.tandfonline.com/page/help/permissions

Notes on Contributors

A. Haroon Akram-Lodhi teaches agrarian political economy at Trent University in Peterborough, Canada. He is also Editor-in-Chief of the *Canadian Journal of Development Studies*.

Alberto Alonso-Fradejas is a PhD candidate at the International Institute of Social Studies (ISS) in the Netherlands, a researcher at the Transnational Institute (TNI) and a Fellow of the Guatemalan Institute of Agrarian and Rural Studies (IDEAR).

Christopher M. Bacon is an Assistant Professor in the Department of Environmental Studies and Sciences, Santa Clara University in California.

Saturnino M. Borras Jr is a Professor of Agrarian Studies at ISS, an Adjunct Professor at the China Agricultural University in Beijing, a Fellow of TNI and a Fellow of the Institute for Food and Development Policy (Food First).

Zoe W. Brent is a PhD student at ISS and a researcher at TNI, Amsterdam.

Jennifer C. Franco is a researcher at the Transnational Institute (TNI) in Amsterdam, and Adjunct Professor at China Agricultural University in Beijing.

Wendy Godek is an independent consultant based in Nicaragua. She is a specialist in food and agriculture systems and sustainable development with a focus on food sovereignty, food security and agroecology.

Todd Holmes is a postdoctoral fellow with the Bill Lane Center for the American West at Stanford University. He earned his PhD in history from Yale University, where for four years he served as the Program Coordinator for the Yale Agrarian Studies Program.

Eric Holt-Giménez has been the Executive Director of Food First since 2006. He is the editor of many academic, magazine and news articles, and lectures internationally on hunger and food system transformation.

Julia is a Presidium member of the Kalimantan Women Alliance for Peace and Gender Justice (AlPeKaJe), West Kalimantan, Indonesia.

NOTES ON CONTRIBUTORS

Tanya M. Kerssen writes and teaches on the political economy of food, agriculture and global development. She currently works with the international research and advocacy organisation, GRAIN.

Giuliano Martiniello is Research Fellow at the Makerere Institute of Social Research, Makerere University, Uganda.

Clara Mi Young Park is a PhD candidate at the ISS in The Hague and a gender officer with the Food and Agriculture Organization of the United Nations (FAO).

Sofía Monsalve Suárez is the Secretary General of FIAN International in Heidelberg, Germany.

Martha Jane Robbins is a PhD researcher at the ISS in the Netherlands and holds a Joseph-Armand Bombardier Canada Graduate Scholarship.

Christina M. Schiavoni is a PhD researcher at the ISS in The Hague.

Louis Thiemann is a PhD researcher at the ISS in The Hague.

Ben White is an Emeritus Professor of Rural Sociology at the ISS in The Hague.

Food sovereignty: convergence and contradictions, conditions and challenges

Alberto Alonso-Fradejas[a], Saturnino M. Borras Jr[a], Todd Holmes[b], Eric Holt-Giménez[c] and Martha Jane Robbins[a]

[a]International Institute for Social Studies (ISS), The Hague, Netherlands; [b]Bill Lane Center for the American West, Stanford University, USA; [c]Institute for Food and Development Policy (Food First), Oakland, USA

This article introduces this special collection on food sovereignty. It frames the collection in relation to a broader political and intellectual initiative that aims to deepen academic discussions on food sovereignty. Building upon previous and parallel initiatives in 'engaged academic research' and following the tradition of 'critical dialogue' among activists and academics, we have identified four key themes – all focusing on the contradictions, dilemmas and challenges confronting future research – that we believe contribute to further advancing the conversation around food sovereignty: (1) dynamics within and between social groups in rural and urban, global North–South contexts; (2) flex crops and commodities, market insertion and long-distance trade; (3) territorial restructuring, land and food sovereignty; and (4) the localisation problematique. We conclude with a glance at the future research challenges at international, national and local scales, as well as at the links between them, while emphasising the continuing relevance of a critical dialogue between food sovereignty activists and engaged scholars.

Introduction

Food sovereignty – as an idea for an alternative food system and as a global social movement – has constantly evolved since its launch by the international agrarian movement, La Via Campesina (LVC), during the period 1993–96. In 2007 a food sovereignty world assembly was held in Mali, where more than 500 advocates coming from 80 different countries gathered for several days to commit themselves, and their respective movements, to the ideals of food sovereignty. The joint declaration from that event became popularly known as the Nyéléni Declaration – a key reference point for the *what, who, why, how, where, when* and *why does it matter* questions in food sovereignty. The name 'Nyéléni' was inspired by 'a legendary Malian peasant woman who farmed and fed her peoples well'.[1] The most abbreviated vision of food sovereignty is captured in the following extended quote from the Nyéléni Declaration:

> Food sovereignty is the right of peoples to healthy and culturally appropriate food produced through ecologically sound and sustainable methods, and their right to define their own food and agriculture systems. It puts the aspirations and needs of those who produce, distribute and consume food at the heart of food systems and policies rather than the demands of markets and corporations. It defends the interests and inclusion of the next generation. It offers a strategy to resist and dismantle the current corporate trade and food regime, and directions for food, farming, pastoral and fisheries systems determined by local producers and users. Food sovereignty prioritizes local and national economies and markets and empowers peasant and family farmer-driven agriculture, artisanal fishing, pastoralist-led grazing, and food production, distribution and consumption based on environmental, social and economic sustainability. Food sovereignty promotes transparent trade that guarantees just incomes to all peoples as well as the rights of consumers to control their food and nutrition. It ensures that the rights to use and manage lands, territories, waters, seeds, livestock and biodiversity are in the hands of those of us who produce food. Food sovereignty implies new social relations free of oppression and inequality between men and women, peoples, racial groups, social and economic classes and generations.[2]

It is clear from the Declaration that food sovereignty is a *political* project. The *how* questions in the political construction of food sovereignty will thus necessarily involve questions about engaging with social forces external to the collective movement that may facilitate or hinder the attainment of food sovereignty. Yet they will also entail a constant, internal renegotiation within the emerging social forces constructing food sovereignty. In the 2007 Nyéléni Declaration, advocates committed thus:

> We are committed to building our collective movement for food sovereignty by forging alliances, supporting each other's struggles and extending our solidarity, strengths, and creativity to peoples all over the world who are committed to food sovereignty. Every struggle, in any part of the world for food sovereignty, is our struggle.

> We have arrived at a number of collective actions to share our vision of food sovereignty with all peoples of this world [...] We will implement these actions in our respective local areas and regions, in our own movements and jointly in solidarity with other movements. We will share our vision and action agenda for food sovereignty with others who are not able to be with us here in Nyéléni so that the spirit of Nyéléni permeates across the world and becomes a powerful force to make food sovereignty a reality for peoples all over the world.

The political build-up from food sovereignty's initial public launch during the 1996 World Food Summit in Rome towards Nyéléni in 2007, as well as the post-2007 momentum in its political construction (idea and social movements), have proven both inspiring and challenging. The initiative has steadily advanced and expanded, albeit unevenly, across space and time. And, in the process, it has answered some questions while also provoking new ones.

A fundamentally contested concept, food sovereignty has – as a political project and campaign, an alternative, a social movement, and an analytical framework – barged into global discourses, both political and academic, over the past two decades. Since then it has inspired and mobilised diverse publics: workers, scholars and public intellectuals; farmers and peasant movements; food vendors and restaurant owners; public health advocates and neighbourhood gardeners; NGOs and human rights activists in the North and global South. The term has become a challenging subject for social science research, and has been interpreted and reinterpreted in various ways by different groups and individuals. Indeed, as it is a concept that is so broadly defined, it spans issues such as food politics, agroecology, land reform, pastoralism, fisheries, biofuels, genetically modified organisms (GMOs), urban gardening, the patenting of life forms, labour migration, the feeding of volatile cities, community initiatives and state policies, public health, climate change, ecological sustainability, and subsistence rights. Similarly the meaning of food sovereignty has morphed and expanded quite significantly beyond the 'rural/agricultural' framing originally given by LVC. Today it stretches across a manifold of socio-political and economic scaffolding: rural/agricultural, rural/non-agricultural, urban/agricultural, urban/non-agricultural, spheres of production, circulation/trade and consumption, the North–South hemispheric divide, state–society institutional spaces, as well as class and other social attributes and identities. Such is a key indication of the relevance – and power – of food sovereignty (FS) as an idea, a social movement, a campaign, and an analytical framework. It also provides us with a glimpse of why FS can be a complicated issue to negotiate within and between social classes and groups across societies.

Since Nyéléni in 2007 FS has significantly gained more ground within the academic community internationally. Slowly FS has been introduced into various academic disciplines, including agrarian political economy, political ecology, international political economy, international relations, ecological economics, world-system studies, social anthropology, development studies, law, sociology, politics, gender studies, public health, and human rights. All the while FS has increasingly garnered the attention of advocates, supporters, sympathisers and sceptics. In the effort to explore what FS means to each of these academic disciplines, and how the academy could contribute to deepening and broadening the conversation around FS, several academics and research institutions organised two international conferences around the theme 'Food Sovereignty – A Critical Dialogue'. The first conference was held at Yale University in September 2013. It was hosted by the Yale Program in Agrarian Studies and the Yale Sustainable Food Project, and coordinated by James C. Scott. Close to 300 participants from around the world attended that conference, where some 82 papers were presented and discussed. The second conference was held four months later, in January 2014, at the International Institute of Social Studies (ISS) in The Hague, and attended by around 350 academics and activists from across Europe. The conferences were co-organised by the Transnational Institute (TNI) and the Institute for Food and Development Policy (Food First). In all, nearly 100 conference papers developed from these two events. Militants, advocates, supporters, and sympathisers – as well as sceptics – from inside and outside the academy gathered at these events and debated in collegial and comradely fashion several critical issues surrounding FS. Indeed, these conferences complemented the earlier, equally critical conferences organised by key scholars including Annette Desmarais, Hannah Wittman, Nettie Wiebe, Harriet Friedmann and Philip McMichael regarding FS, and proved to be immensely productive, both politically and academically. The editors of this special collection were among the organisers of these two conferences.

Three journal special issues have been produced from the said conferences. The first collection appeared in the *Journal of Peasant Studies*, 41, no. 6 (2014), edited by Marc Edelman, James C Scott, Amita Baviskar, Saturnino M Borras Jr, Deniz Kandiyoti, Eric Holt-Giménez, Tony Weis and Wendy Wolford. This collection focuses on the agrarian dimensions of FS. The second collection is forthcoming in a special issue of *Globalizations* (summer 2015), edited by Annie Shattuck, Christina Schiavoni, and Zoe VanGelder. It focuses on general globalisation-related issues and includes a number of contributions linked to issues in the global North. The third collection is this current special issue of *Third World Quarterly* – the content and focus of which we will introduce below.[3]

Building on the critical dialogue at Yale and ISS, this present collection identifies a number of key questions regarding FS. What does (re)localisation mean? Although the concept stands at the centre of the food sovereignty narrative, (re)localisation has rarely been problematised systematically in any academic terms. How does the notion of food sovereignty connect with similar and/or overlapping ideas historically? How does it address questions of both market and non-market forces in a dominantly capitalist world? There is a tendency in the food sovereignty narrative to sidestep divisive issues such as gender: how does FS deal with such differentiating social contradictions? The alternative FS both embodies and promotes often focuses on scattered localised food systems. But how does the movement deal with larger issues of nation-state, where a largely urbanised world of non-food producing consumers harbours interests distinct from those of farmers? How does food sovereignty address the current trends of crop booms, as well as other alternatives that do not sit comfortably within the basic tenets of food sovereignty, such as corporate-captured fair trade? How does FS grapple with the land question and move beyond the narrow 'rural/agricultural' framework? These are among the current questions facing food sovereignty (also see the key questions raised in the *Journal of Peasant Studies* and *Globalizations* special issues). Such questions, indeed, call for a new era of research into FS, a movement and theme that in recent years has inspired and mobilised tens of thousands of activists and academics around the world: young and old, men and women, rural and urban. In the remainder of this article we will elaborate on some of these questions.

Dynamics within and between the rural–urban, North–South divides
Farming and non-farming rural social groups
Amid the far-reaching promises of food sovereignty questions still abound regarding where non-food producers in the rural areas of both the global North

and global South fit, intellectually and politically, within the movement's future trajectory. In 1939 Carey McWilliams' *Factory in the Fields* put forth the earliest critique of the industrial agriculture that was taking root in the USA and, more importantly, of the impoverished and racialised caste of farm workers that served the emerging order.[4] Seventy-five years later McWilliams' work continues to prove prescient. Just as the now-familiar features of capitalist agriculture – corporate control, land concentration, and increasingly biotechnological industrial inputs and processing – have spread throughout the global North, so too has the system of cheap, racialised foreign agricultural labour that has long underpinned it. And while campaigns for farm worker unionisation and social justice sporadically swept through the USA and parts of Europe between the 1930s and 1980s, such concern for the rural working poor in the North has waned amid the rise of food politics, where method and place of production, from organic and sustainable to slow food and buy local movements, has obfuscated the exploitation and plight suffered by agricultural labour.[5]

Indeed, food sovereignty, as both a global movement and an intellectual field of study, affords the opportunity to refract important light on rural labour as well as advance discussion on how to bridge the class, racial and ethnic divides that have long impeded collective action and consciousness in the rural North. This is certainly no easy task. As a number of social scientists have noted, the racial minorities and foreign immigrants working the industrial assembly lines of slaughterhouses and processing plants in North America employ a range of workplace logics to formulate quasi-class identities to differentiate themselves not only from those occupying lower status jobs at the industrial plant, but especially from those toiling in the agricultural fields.[6] At the same time landless contract farmers – a common feature in the corporate landscape of the global North, who are essentially reduced to selling their labour power – differentiate themselves from rural workers through a class identity, which is significantly reified by race and ethnicity. In all, these barriers underscore how distancing in industrial agriculture operates to sever connections among rural workers as often as between producer and consumer.[7]

In the global South such barriers to collective action and consciousness among rural workers seem even more daunting for food sovereignty. As some of the contributors to this special issue explore in detail, 're-peasantisation' not only risks invoking an agrarian imaginary that, for some, falls far outside the realm of realistic possibility in the twenty-first century, but also homogenises the thicket of social, economic and cultural complexities afflicting the rural communities of the South.[8] Class identity – and the varying ideologies, concerns and political allegiances inherent in it – fuels conflict and antagonism between landless labourers, small farmers and petty commodity processors, just as the wide range of economic sensibilities among rural peoples sow additional seeds of fragmentation. Cultural tensions further aggravate such divides, especially for ethnic minorities and women, the latter increasingly accounting for the majority of farm workers in many regions.[9] Taken together, these fissures within the rural communities of the global South spotlight the barriers to the 'radical egalitarianism' promised by food sovereignty, as well as the mixed appeal – and relatively limited possibility – of the movement's smallholder, peasant idyll. Where do landless farm workers, especially ethnic minorities and women, fit within this agenda of agrarian change?

Why do some labourers prefer engagement with corporate agriculture to small-scale farming? And what erodes the common ground of solidarity in rural reform movements, from Latin America to Africa to Southeast Asia? These are some of the most sensitive and difficult political questions confronting FS activists and some of the challenging academic research questions that several scholars have also identified.[10] Other persistent contradictions and tensions are those between sedentary farmers and pastoralists, agriculturalists and indigenous peoples, and rich farmers and poor farmers, among others. It is not that FS activists have not spotted these complicated dimensions of the movement-building process – indeed they have. Yet, contrary to sweeping claims and movement slogans, these are political dilemmas not easily resolved.

Tension in producer–consumer, rural–urban links

Food sovereignty's original social base is located in the peasantry of the global South and the small-scale, family farm sector of the global North. Because it is one of the few broad political platforms today globally contesting neoliberal capitalism, food sovereignty has spread across food system struggles to urban and peri-urban areas of the global North, where students, socially conscious consumers, farm and food workers and food justice advocates have embraced it as a banner for social justice and food system transformation.[11]

At first glance food sovereignty in the South is primarily rural and peasant-based, while its expression in the North is largely (though not solely) urban and consumer-based – particularly in North America (where there are key farm movements around FS too). While this broadly mirrors the rural–urban demographic distributions of industrialised and less industrialised countries that characterised the 'development decades', the de-industrialisation of large areas in the North and the rapid industrialisation of some boom economies in the South (more specifically, Brazil, Russia, India, China and South Africa (BRICS) and some key middle-income countries) demands a nuanced understanding of the overlaps, exceptions and contradictions among these actors and settings, as well as analyses from different disciplines and multiple vantage points. The agrarian transitions of capitalist agriculture are part of a longer and larger transition to a globally integrated capitalist food system characterised by the political influence of multilateral institutions, liberalised trade, transnational oligopolies, global value chains, 'supermarket-isation' and the grain–oilseed–livestock complex.[12] These structural changes have been central to a number of global-scale crises (such as species extinction and global climate change), disproportionately affecting vulnerable populations and communities around the world, that have also driven the expanding agenda of food sovereignty.

The rise of food sovereignty as a demand signals a significant political shift toward recognising the importance of agrarian organisation in the spheres of both production and consumption. The particular ways in which food sovereignty is incorporated into struggles around food, land, labour and environment in the food systems of industrialised countries depends only partly on how the term has been defined by LVC, the participants at the Nyéléni food sovereignty world assembly (or agrarian sociologists). The fact that so many people and groups are taking up the FS call without consulting with LVC or agrarian sociologists is, after all, what gives FS its political power. How the global processes of industrialisation, liberalisation and financialisation currently affect people's food systems plays an equally critical role in the process of FS becoming incorporated into these political struggles. The multiple combinations of factors and contexts – the mix of agricultural biotech industrialisation and market liberalisation; North–South and rural–urban dynamics – are leading to variation, overlap, unevenness and contradictions among and within the communities and social movements struggling for dignified livelihoods and healthy, sustainable food systems. These are further complicated by layers of vulnerability rooted in diverse social attributes like gender, race, caste, nationality, religion and ethnicity, as well as localised conflicts over water, land, jobs and gentrification.

Given these complicated and volatile scenarios, one wonders how a single concept like food sovereignty could have any practical political meaning. Nevertheless, clear linkages exist between the dispossession of peasant and indigenous communities in the global South and the epidemic of diet-related diseases in low-income communities of colour in the USA. And similar links are evident between the oppression of African immigrant farmworkers in the fruit and vegetable farms of southern Europe and their Latino counterparts in North America. Yet while these relationships may be understood by using the food sovereignty lens in the abstract, it is less than clear how FS will bring social movements working under such disparate conditions into a common programme for transformation.

In the global North food sovereignty advocates face tensions and contradictions arising from more reformist approaches from government, the private corporate sector and big philanthropy. Such approaches, in the best of cases, advocate food security without structural changes, which have often led to nothing more than the 'mainstreaming' of organic food or Fair Trade products within monopoly firms whose business model is devastating small and mid-sized family farms. The political impossibility of introducing significant change into the US Farm Bill or the European Common Agricultural Policy has driven food sovereignty's focus to the spheres of local communities and local government. Here advocates might change some local policies and practices, but there are relatively few resources with which to build lasting, transformative institutions or social structures. The lack of public resources drives some factions of the food sovereignty movement towards the non-profit sector, where philanthropic foundations controlling the purse strings tend to depoliticise agendas of social change, diluting both their class power and capacity for coalition building.[13]

Food sovereignty in the global North is clearly both strengthened and complicated by the broad ways in which the concept has been incorporated across food system struggles.[14] Understanding the obstacles to and opportunities for convergence and the re-politicisation of the food movement are perhaps the most immediate political challenges facing the food sovereignty agenda in the global North.[15] Further enquiry by researchers in agrarian studies, food systems and social movements will help to shed light on these developments.

Critical gender perspectives

Food sovereignty includes a call for gender equality within a fundamental reorganisation of the food system.[16] LVC, a key promoter of FS, has stood at the forefront of advancing such egalitarianism by institutionalising gender balance in its leadership structures and making space for women's voices (especially through a dedicated Women's Commission). Moreover, the organisation has launched a global campaign to stop violence against women, a campaign that goes beyond domestic violence to include 'the structural violence that women have to confront each day and that has been systematically silenced, made to appear natural, and rendered invisible by patriarchal capitalist society'.[17]

The adoption of FS by the World March of Women (WMW), and their political alliance with LVC is an indication of FS's deep potential for addressing inequities of gender. By incorporating FS into their demands, the WMW not only broadened and deepened the voice of rural women in their political campaign for women's rights; they were able to affect political agendas far beyond the women's movement. As Miriam Nobre of the WMW points out:

> We united behind the principles of food sovereignty first because our rural sisters in the World March of Women invited us to join their struggle for land and fair conditions to live and produce as farmers...We also understand that food sovereignty allows us to expand the feminist movement's horizons...Our contribution as a feminist movement is to link the goal of women's autonomy with the vision of sovereignty for all people.[18]

As Park et al note in this issue, food sovereignty – in its reorganisation of power dynamics within the food system – is well placed to explore gender dimensions. Yet they also caution that the unifying discourses employed by FS movements in advocacy and mobilisation campaigns blur important differences, such as experience and class, leading to a tendency towards homogenisation that glosses over gender dimensions. Along these lines Agarwal has critiqued what she views as an important contradiction within FS: advocating gender equality, on one hand, while simultaneously promoting the 'family farm' idyll, on the other.[19] She argues that:

> More particularly, family farms do not provide autonomy to women workers or the means to realize their potential as farmers. Hence a nod toward gender equality is not enough. The problems women face as farmers are structural and deep-rooted, and would need to be addressed specifically. This would include redistributing productive assets such as land and inputs within peasant households in gender-equal ways, and directing state services to cater better to the needs of women farmers, such as services relating to credit, extension, training, information on new technology, field trials, input supply, storage and marketing. (Ibid., 1255)

Park et al follow this logic of specificity and examine the gender dimension in various case studies regarding land rights, division of labour and access to employment. They make an important call for FS to 'address gender inequalities systematically as a strategic element in its construct and not only as a mobilising ideology'. This distinction proves equally pertinent for both FS actors and those engaging in the analysis of food sovereignty discourse and practice.

Flex crops and commodities, market insertion and long-distance trade

There is nothing in the Nyéléni Declaration that precludes long-distance trade. There are, however, two related issues that bring us back to the issue of trade, namely the explicit position of FS against food dumping, whether through official food aid or at prices below production costs, and the call for the 're-localisation' of the food system.[20] In turn, these drag us back to an old agrarian political economy discussion that has recently been resurrected within mainstream agricultural economics by new institutional economists. The discussion also seems to have found renewed expression in various NGO mantras, including 'corporate social responsibility', 'fair trade', 'business and human rights', 'sustainable rounds table initiatives' and 'insertion into the value chain'. In its 2008 *World Development Report*, the World Bank claimed that 'it is time to place agriculture afresh at the centre of the development agenda'.[21] In so doing, the World Bank envisages an agricultural system 'led by private entrepreneurs in extensive value chains linking producers to consumers and including many entrepreneurial smallholders supported by their organizations'.[22] This advocacy received a boost from what Borras et al call the rise of 'flex crops and commodities' – crops and commodities that have multiple uses and can be flexibly interchanged across food, feed, fuel, fibre and other sectors.[23] This complex has led to a recent transition from the conventional 'value-chain' to what Virchow et al call a 'value web' – the interlocking of various value chains.[24]

The rise of flex crops and commodities, often embedded in complex value webs, do not inherently mean something adverse for small farmers or for the environment. But this growing phenomenon is a new context for many FS front-liners, which in turn spurs questions regarding long-distance trade, free trade agreements and international regulations and standards, and, perhaps most importantly, the placement of small farms within this value web in ways that are significantly different from Jan Douwe van der Ploeg's idea of entrepreneurial 'new peasantries'.[25] Three relevant political processes are important for the further examination of the challenges they impose upon FS political construction: (1) international/multilateral negotiations on trade, investment and trade-related intellectual property issues, among others, in the World Trade Organization (WTO), as discussed by Akram-Lodhi and Bacon in this issue; (2) regional multilateral/bilateral negotiations via free trade agreements, especially after WTO negotiations broke down in the Doha Round because of the opposition of BRICS and many middle-income countries to the protectionist measures long enjoyed by Northern countries; and (3) the proliferation of various regulatory institutions and standards whose impact on poorer producers and consumers is not always palpable and positive. Examples of this third process can be seen in the social, labour and environmental commodity certification schemes exercised by private bodies like the Roundtable on Sustainable Palm Oil, and in fair trade (corporate or otherwise), both of which circumvent the regulatory power of the nation-state.

LVC framed the original formulation of FS in the context of a struggle against neoliberal globalisation manifested in unbalanced international terms of trade negotiated under the Uruguay Round of the General Agreement on Tariffs and Trade and later under the WTO. The objective conditions of agricultural production, trade and consumption have since then become even more complicated, partly through the rise of flex crops and commodities and the reorganisation of production, fragmentation, industrialisation, circulation and consumption through their related value webs. How then can FS be repositioned within this changing context? What does this recent phenomenon imply for FS constituencies? The answers to these questions are not obvious. Nonetheless, from the perspective of the mode of agricultural production envisaged by FS, the 'terms' of inclusion in the flex crop value webs, ranging from less to more favourable,[26] are of greater relevance. The politics of inclusion/exclusion in/from capitalist agriculture as producers and/or as workers are discussed by Martiniello in the case of Acholi farmers in Northern Uganda, by Bacon in the case of fair trade smallholder coffee producers in Nicaragua, and by Kerssen in the case of indigenous quinoa farmers in the western highlands of Bolivia (all in this collection). Thiemann brings the discussion to the question of agricultural investment; he challenges the assumption that 'corporate investment' is the only game in town.[27] And in her case study of cacao growers in upland Sulawesi, Indonesia, Tania Li also reminds us that 'even when small-scale farmers are untouched by land grabbing or corporate schemes, as in this case, expanding their capacity to exercise control over their food, their farms and their futures is still a huge challenge'.[28]

There are two important insights that the cases above lay bare. On one hand, the peasantry across the South–North hemispheric divide engage differently with the market and corresponding regulatory institutions. Thus, the notion of socially differentiated producers is a key analytical lens that remains relevant in the study of FS, as consistently discussed in this paper and by Edelman et al. and Bernstein, among others.[29] This would include class and other social attributes and identities, such as race and ethnicity, as well as gender. On the other hand, FS needs to be more explicit about its idea of long-distance trade, and the conditions of trade that could prove beneficial for small-scale producers, family farmers and working people. Burnett and Murphy represent some of the first scholars to systematically address this important and controversial issue.[30]

Territorial restructuring, land and food sovereignty

Calls for food sovereignty must address the question of land in ways that fully capture the FS scope. Historically the commodification of land under capitalism has affected all forms of land access, use and tenure. Food sovereignty evolved in no small measure in response to capitalist agriculture's three-decade neoliberal trend of market-driven dispossession of peasant lands, reflected in LVC's longstanding demands for state-led, redistributive land reform. More recently, large-scale corporate land grabs and the financialisation of agricultural land have brought forth new calls for 'democratic land control by working people and peoples', framed as 'land sovereignty' in this issue by Borras et al.

The emergence of land sovereignty as a political demand reflects two conceptual–geographic shifts: from farm system to food system, and from farm scale to territorial scale. These two shifts originate in separate, but converging, loci. On one hand, agrarian capital and other extractive industries are transforming ever-larger swaths of forest, wetlands and agricultural land, driving hundreds of thousands of rural people from the countryside, or trapping them in adverse incorporation schemes. This has widened the scale of resistance among affected peasantries, family farmers, fisher-folk, pastoralists and indigenous communities as they fight for their respective livelihoods – increasingly in shared territories.[31]

On the other hand, the convergence of consumers, peasants, family farmers, fisher-folk, pastoralists and indigenous communities in the struggle against capitalist enclosures of land, food and resources has produced a widening frame of resistance.[32] Urban, peri-urban and rural lands are being redefined by production and export corridors. The expansion of flex crops and commodities are steadily remaking food sheds, linking them to global value webs, in which ever-growing supermarket networks, modern grain–oilseed–livestock complexes, and emerging biochemical complexes exploiting the multiple uses of agricultural and forest biomass, are deeply reshaping the meaning and workings of 'agriculture' in contemporary capitalism. Ancient food crops like teff, quinoa and maca are rapidly being commodified as global speciality products, pricing them out of local markets and turning diverse, poly-cultural landscapes into large-scale monocultures. In this process other key resources – seeds, water and forests, but also knowledge – are redefined and appropriated in the service of capital accumulation.[33]

The academy is also going through an intellectual process of reframing and rescaling. Food systems studies have popped up on university campuses around the world. Some of the largest of these programmes are, ironically, financed by the very companies presently expanding their operations into sub-Saharan Africa, Asia, the former Soviet Union and Latin America. Agro-ecology – a multidisciplinary science that has largely co-evolved with food sovereignty – has steadily broadened its scope from farm, to watershed, to food-system scales, meeting stiff resistance from industrially financed science, as well as grudging recognition from the multilateral institutions tasked with agricultural development and food security.[34]

While arenas of resistance are found across the spectrum of institutional scales – from food policy councils in Northern cities and 'citizen juries' in Southern rural communities, to the Committee on World Food Security – the importance of 'grounding' resistance while integrating multiple actors at multiple scales has never been greater, or more challenging. The call for land sovereignty and the democratic control over land and territory responds to capitalism's regressive processes of 'territorial restructuring'. Territorial restructuring seeks control over the places and spaces where surplus is produced by shaping and controlling the institutions and social relations that govern production, extraction and accumulation. Capital exerts this control over specific territories through development banks, private firms and national governments, and by using other global and local institutions and organisations. The accumulated result of these activities, tensions, and alliances between different actors is the regressive restructuring of territorial spaces and places, such as the markets, municipalities, farms, forests and roads that make up local institutions and landscapes.[35] Territorial restructuring encounters friction, slippage and resistance, all of which can lead to unexpected outcomes for agrarian and other extractive capital. Effective resistance in favour of indigenous and peasant livelihoods or redistributive land and water policies requires not only identifying the inherent fissures of territorial restructuring but political mobilization for redistributive and democratic restructuring on a territorial scale. This is the essence of land sovereignty. Whether such broadening of the 'land' framework within food sovereignty is politically feasible and, if so, what tensions and synergies it will produce is another critical area for future research.

Localisation problematique

Food sovereignty includes a central call for the localisation of the food system. This is in part a reaction to, and a way to subvert, the various types of 'distancing' that the current industrial food system has created. However, localisation discourse is often vague, theoretically or politically assumed rather than empirically examined or specified. As Robbins explains in this issue, 'the ambiguous nature of defining "local" and "local food systems" makes this task more challenging as they are not only defined in geographic terms but also by "social and supply chain characteristics"'. It is one of the key dimensions of FS that has remained underexplored. It should be on the agenda of any future critical dialogue regarding FS.

Various authors in this collection examine the localisation narrative of FS in different contexts and to varying extents. This stands as one of the important contributions the current collection offers to the literature and analysis of FS. Robbins addresses this question comprehensively. She examines the distancing–localisation problematique from five interrelated perspectives: production and consumption, distant markets, peasant lands, rural and urban, and agriculture and nature. In offering a schematic to categorise local food practices by their method, character and scale, she notes that food sovereignty requires more than just small-scale production or local markets. Moreover, Robbins observes that many local food initiatives neither fit neatly into the food sovereignty framework nor resemble mere extensions of the capitalist industrial model. Instead, most local initiatives fall somewhere in between the two poles.

The issue of scale is of particular interest at the intersection between localisation and food sovereignty. Can local food systems adequately feed a burgeoning global population? What does scaling up local food look like and is it possible to achieve without losing the ecological and social connectedness that food sovereignty promotes? Many of the authors in the current collection grapple with this conundrum in various ways. Bacon outlines the case that fair trade can be in line with food sovereignty, and through his dissection of the divergence between 'Fair Trade' governing bodies, illustrates the perils of 'mainstreaming' or scaling up an alternative approach. He highlights how the defining features of the alternative tend to erode as the consumer base grows. Thiemann offers a theory of how to move between what he terms the 'layers' of FS discourse – the abstract layer of generalised principles and stances, and the concrete layer of implemented ideas – using what he terms an investment lens. He proposes that by fleshing out gaps in the FS framework, the potential of FS will become clearer. What localisation within a food sovereignty frame looks like and how local food systems can radiate outwards to reach a large number of consumers while addressing questions of accessibility, affordability and ecological sustainability is one of those gaps. Brent et al outline the challenge of maintaining affordability for low-income urban consumers, while simultaneously maintaining decent prices for struggling producers, without engaging in depoliticising schemes such as corporate sponsorship. Iles and Montenegro's concept of 'relational scale' is an important contribution to bringing the critical dialogue on the localisation problematique forward.[36]

Continuing a critical dialogue at various levels between activists and academics

Moving forward, food sovereignty stands as a significant and fertile field for academic research. As many of the contributors in this collection underscore, such research needs to advance a multi-lens view and examination of FS, a comprehensive approach that grapples with the cultural, economic and political complexities that confront the movement at the international, national and local levels. Such explorations will not only help further develop future intellectual and political pathways for food sovereignty, but also better situate academics, policy makers and citizens to more effectively address political questions in global agriculture in the twenty-first century.

Neoliberalism has increasingly become the all-encompassing political catchword in critiques of global capitalism and the facets of international political economy that underpin it – and, in view of international trade policies and development schemes, for good reason. Yet in contemporary agriculture the term can obfuscate more than it highlights, especially in the global North where the term 'neoliberalism' has come almost to resemble a rhetorical act of group think rather than empirical analysis. For well over a century agriculture in the North has thrived upon an ever-expanding array of state subsidies, policies that by any stretch of the definition do not fit the neoliberal ideal popularly ascribed to the political likes of Ronald Reagan and Margaret Thatcher. Here scholars of the global North need to forward a stringent delineation of political economy between domestic production and trade policy, and explore the various political and economic mechanisms that have allowed this chasm of contradiction to continue between the North and South. Moreover, such examinations would help address the international impediments FS confronts, and advance more revealing and in-depth studies on the hindering roles and policies of global institutions such as the WTO, IMF and World Bank, whose programmes have long sowed the seeds of financial and food dependency.

Future research also needs to contextualise the changing global context food sovereignty increasingly confronts. The converging crises of food, fuel, energy, climate and finance, as well as the development of the flex crop and commodity complexes that have significantly contributed to the rise of newer hubs of global capital, such as BRICS and some middle-income countries, have significantly reshaped international political economy. How has this change affected local (agrarian/food) political economies? Moreover, how have the instruments of international governance been transformed as a result of these recent phenomena? That these developments are all largely linked to the global food system highlight additional facets of future food sovereignty research.

At the national level future research in the global South needs to more thoroughly engage with regional politics and navigate its complexities to better reveal the areas in which the nation-state can hinder and facilitate democratic control over key resources: land, water, forest, seeds. For scholars and FS advocates alike this line of study would not only entail a deconstruction of the state – an entity all too often homogenised – but would also involve grappling with what is ideal and what is possible in regard to reform, as Godek has attempted to do in this issue in the case of Nicaragua. How reform agendas have fared in the halls of government stands as a rich example. To date, FS legislation has been passed in Mali, Venezuela, Senegal, Nepal, Bolivia, Ecuador, Nicaragua and the Dominican Republic. Comparative examinations of these (partial) political victories, as well as of those efforts thwarted, would greatly further analysis and understanding of the successful strategies, activism, narratives and imagery employed in the respective regions. Other areas of research relating to the state and resource/territorial control would include political representation, access to public and private land, economic and trade policy at both state and local levels, sustainable water and infrastructure development, as well as seed and other farm input control. Building upon both state and international analyses, future research could also help provide a better understanding of the political-economic linkages between farm, statehouse and market and, in so doing, unpack the complex interrelationship that exists in some regions between food sovereignty, food security and international trade.

At the local level future research needs similarly to address the complexities of rural communities and their urban and peri-urban neighbours. Here again comparative studies would not only greatly enhance our understanding of how food sovereignty is expressed and interpreted on the ground, but also highlight the various impediments to democratic resource control that persist at local levels. As some of the contributors to this collection note, FS cannot be exported in a blanket manner but needs to adapt to the political, social, and cultural rhythms of local peoples. Exploring these rhythms at the ground level would help foster such local and regional adaptations, and resituate scholarly attention toward the praxis of rural communities. Within this vein, future research needs to more thoroughly examine the social and cultural constructs of hierarchies within a given region and community, such as ethnicity, gender and class-identities, and how these constructs pervade local institutions and impede, or not, collective action for reform. Moreover, grappling with these cultural and economic complexities would also correct the homogenised – if not idealised – view of the 'peasant'. Indeed, this correction would deepen our understanding not only of the blurring line between consumer and producer in the global South, and the grey zone between market and peasant economy in which many rural people reside, but also the preference for farm labour over small farming among segments of the rural population in the South.

How the political dynamic in each of these levels reshape one another is a critical process to pay closer attention to in the future. How to carry out rigorous research that takes these linkages between levels into account is an urgent and important challenge. Iles and Montenegro have offered useful ways on how we can think about these questions.[37]

We close by returning to the declaration of principles and strategies for FS political construction that opened this discussion. The Nyéléni Declaration is a call to action for all peoples to unite and overthrow the currently dominant food system with an alternative, namely food sovereignty. It is a call for a pluralist, broad-based, cross-class alliance spanning intra- and inter-social class and other attributes and identities, rural and urban, and North–South divides. An anti-systemic campaign for a comprehensive alternative food system has to be simplified, direct, clear and easy to grasp. In this context the principles and slogans expressed in the Nyéléni Declaration fit the bill in a powerful way. Yet reality often proves much more complicated. Social relations between classes and groups who have a stake in developing an alternative food system can be messy, with socioeconomic interests and political standpoints often spurring competition and antagonism.

There are twin tasks for activists and academics. On one hand, confronting – rather than backing away from – these contradictions and dilemmas will be a necessary political task for all food sovereignty advocates. What activists cannot afford to do is to stand idle because of the seemingly irreconcilable contradictions and dilemmas they face. On the other hand, researching these contradictions and dilemmas in acute ways that raise difficult, but helpful, questions will be a vital contribution from engaged academics. What scholars cannot afford to do is to avoid the food sovereignty issue, or heckle it from afar without fully appreciating the messy contradictions and complexities that ensnarl the movement. The present collection is a modest but important contribution to these twin tasks.

Notes

1. "Declaration of Nyéléni," 2007.
2. Ibid.
3. In addition, the conference organisers also managed to produce high-quality video clips of the plenary speakers at the conferences. These video clips are available on the websites of the conference co-organisers (Yale Program in Agrarian Studies, Food First and TNI) as well as on YouTube. It is important to read the individual contributions in these collections in the context of the broader critical dialogue launched at Yale and ISS. Moreover, re-reading other key academic texts on FS, such as Claeys, *Human Rights*; Wittman, Desmarais, and Wiebe, *Food Sovereignty*; and Patel "Food Sovereignty", is also encouraged.
4. McWilliams, *Factories in the Field*.
5. Majka and Majka, *Farm Workers*; and Mooney and Majka, *Farmers' and Farm Workers' Movements*. On food politics, see Pollan, *Omnivore's Dilemma*; Pollan, *In Defense of Food*; and Weber, *Food Inc.* For an important exception, see Guthman, *Agrarian Dreams*.
6. Pachirat, *Every Twelve Seconds*; and Striffler, *Chicken*.
7. Aronowitz, *The Politics of Identity*; and Pachirat, *Every Twelve Seconds*.
8. Bernstein, "Food Sovereignty via the 'Peasant Way'"; and Holmes, "Farmers Market."
9. Jalali and Lipset, "Racial and Ethnic Conflicts"; and Mills, "Gender and Inequality."
10. See, for example, Bernstein, "Food Sovereignty via the 'Peasant Way'"; and Agarwal, "Food Sovereignty."
11. Holt-Giménez, *Food Movements Unite*; and Desmarais and Wittman, "Farmers, Foodies and First Nations."
12. McMichael, "A Food Regime Genealogy."
13. See Brent et al. in this collection
14. Ibid.
15. Amin, "Food Sovereignty."
16. Patel, "Food Sovereignty."
17. LVC, "La Via Campesina."
18. Nobre, "Women's Autonomy," 303.
19. Agarwal, "Food Sovereignty."
20. See Robbins in this issue.
21. World Bank, *World Development Report*, 1.
22. Ibid., 8.
23. Borras et al., "Towards Understanding." On flex crops and commodities, see the discussion forum in the *Journal of Peasant Studies* (2015) and TNI's "Think Piece Series on Flex Crops & Commodities". http://www.tni.org/category/series/think-piece-series-flex-crops-commodities.
24. Virchow et al., "The Biomass-Based Value Web as a Novel Perspective on the Increasingly Complex African Agro-Food Sector."
25. Ploeg, *The New Peasantries*.
26. Du Toit, "Social Exclusion Discourse."
27. Se also Kay, *Positive Investment Alternatives*; Kay, *Policy Shift*; and HLPE, *Investing in Smallholder Agriculture*.
28. Li, "Can there be Food Sovereignty Here?", 6.
29. Edelman et al., "Introduction"; and Bernstein, "Food Sovereignty via the 'Peasant Way'."
30. Burnett and Murphy, "What Place for International Trade?"
31. Edelman et al., "Global Land Grabs"; Wolford et al., "Governing Global Land Deals"; and White et al., "The New Enclosures."
32. Margulis et al., "Land Grabbing and Global Governance"; and Borras and Franco, "Global Land Grabbing."
33. On water grabbing, see, for example, Franco et al., "The Global Politics."
34. Holt-Giménez and Altieri, "Agroecology."
35. Holt-Giménez, *Land, Gold, Reform*.
36. Iles and Montenegro, "Sovereignty at what Scale?"
37. Ibid.

Bibliography

Agarwal, B. "Food Sovereignty, Food Security and Democratic Choice: Critical Contradictions, Difficult Conciliations." *Journal of Peasant Studies* 41, no. 6 (2014): 1247–1268.
Amin, S. "Food Sovereignty: A Struggle for Convergence in Diversity." In *Food Movements Unite! Strategies to Transform our Food Systems*, xi–xviii. Oakland, CA: Food First Books, 2011.
Aronowitz, S. *The Politics of Identity: Class, Culture, Social Movements*. London: Routledge, Chapman and Hall, 1992.

Bernstein, H. "Food Sovereignty via the 'Peasant Way': A Sceptical View." *Journal of Peasant Studies* 41, no. 6 (2014): 1031–1063.

Borras Jr., S. M., and J. C. Franco. "Global Land Grabbing and Political Reactions 'from Below'." *Third World Quarterly* 34, no. 9 (2013): 1723–1747.

Borras Jr., S., J. C. Franco, R. Isakson, L. Levidow, and P. Vervest. *Towards Understanding the Politics of Flex Crops and Commodities: Implications for Research and Policy Advocacy.* Think Piece Series on Flex Crops and Commodities No. 1. Amsterdam: Transnational Institute, 2014.

Burnett, K., and S. Murphy. "What Place for International Trade in Food Sovereignty?" *Journal of Peasant Studies* 31, no. 6 (2014): 1065–1084.

Claeys, P. *Human Rights and the Food Sovereignty Movement: Reclaiming Control.* London: Routledge, 2015. 'Declaration of Nyéléni'. February 27, 2007. http://www.nyeleni.org/IMG/pdf/DeclNyeleni-en.pdf.

Desmarais, A. A., and H. Wittman. "Farmers, Foodies and First Nations: Getting to Food Sovereignty in Canada." *Journal of Peasant Studies* 41, no. 6 (2014): 1153–1173.

Du Toit, A. "Social Exclusion Discourse and Chronic Poverty: A South African Case Study." *Development and Change* 35, no. 5 (2004): 987–1010.

Edelman, M., J. C. Scott, A. Baviskar, S. M. Borras Jr., D. Kandiyoti, E. Holt-Giménez, T. Weis, and W. Wolford. "Introduction: Critical Perspectives on Food Sovereignty." *Journal of Peasant Studies* 41, no. 6 (2014): 911–931.

Edelman, M., C. Oya, and S. M. Borras Jr. "Global Land Grabs: Historical Processes, Theoretical and Methodological Implications and Current Trajectories." *Third World Quarterly* 34, no. 9 (2013): 1517–1531.

Franco, J. C., L. Mehta, and G. J. Veldwisch. "The Global Politics of Water Grabbing." *Third World Quarterly* 34, no. 9 (2013): 1651–1675.

Guthman, J. *Agrarian Dreams: The Paradox of Organic Farming in California.* Berkeley: University of California Press, 2004.

High Level Panel of Experts on Food Security and Nutrition (HLPE). *Investing in Smallholder Agriculture for Food Security.* Rome: Committee on World Food Security, 2013.

Holmes, T. "Farmers' Market: Agribusiness and the Agrarian Imaginary in California and the Far West." *California History* 90, no. 2 (2013): 24–41.

Holt-Giménez, E., ed. *Food Movements Unite! Strategies to transform the Food System.* Oakland, CA: Food First Books, 2011.

Holt-Giménez, E. *Land, Gold, Reform: The Territorial Restructuring of Guatemala's Highlands.* Development Report 16. Oakland, CA: Institute for Food and Development Policy, 2007.

Holt-Giménez, E., and M. A. Altieri. "Agroecology, Food Sovereignty, and the New Green Revolution." *Agroecology and Sustainable Food Systems* 37, no. 1 (2013): 90–102.

Iles, A., and M. Montenegro de Wit. "Sovereignty at what Scale? An Inquiry into Multiple Dimensions of Food Sovereignty." *Globalizations* ahead-of-print (2014): 1–17.

Jalali, T., and S. M. Lipset. "Racial and Ethnic Conflicts: A Global Perspective." *Political Science Quarterly* 107, no. 4 (1992): 585–606.

Kay, S. *Policy Shift: Investing in Agricultural Alternatives.* Amsterdam: Transnational Institute, 2014.

Kay, S. *Positive Investment Alternatives to Large-scale Land Acquisitions or Leases.* Amsterdam: Transnational Institute, 2012.

La Via Campesina (LVC). "La Via Campesina: Day of Action against Violence Towards Women." Press Release, November 25, 2014. http://viacampesina.org/en/index.php/main-issues-mainmenu-27/women-main menu-39/1702-la-via-campesina-day-of-action-against-violence-towards-women.

Li, T. M. "Can there be Food Sovereignty Here?" *Journal of Peasant Studies* (2014): doi:10.1080/03066150.2014.938058.

Majka, L., and T. Majka. *Farm Workers, Agribusiness, and the State.* Philadelphia, PA: Temple University Press, 1982.

Margulis, M. E., N. McKeon, and S. M. Borras Jr. "Land Grabbing and Global Governance: Critical Perspectives." *Globalizations* 10, no. 1 (2013): 1–23.

McMichael, P. "A Food Regime Genealogy." *Journal of Peasant Studies* 36, no. 1 (2009): 139–169.

McWilliams, C. *Factories in the Field: The Story of Migrant Farm Labor in California.* Boston, MA: Little, Brown, 1939.

Mills, M. B. "Gender and Inequality in the Global Labor Force." *Annual Review of Anthropology* 23 (2003): 41–62.

Mooney, P., and T. J. Majka. *Farmers' and Farm Workers' Movements: Social Protest in American Agriculture.* New York: Twayne Publishers, 1995.

Miriam Nobre. "Women's Autonomy and Food Sovereignty." In *Food Movements Unite.*, edited by Eric Holt Giménez, 293–306. Oakland, CA: Food First Books, 2011.

Pachirat, T. *Every Twelve Seconds: Industrialized Slaughter and the Politics of Sight.* New Haven, CT: Yale University Press, 2011.

Patel, R. "Food Sovereignty." *Journal of Peasant Studies* 36, no. 3 (2009): 663–706.

Ploeg, J. D. van der. *The New Peasantries: Struggles for Autonomy and Sustainability in an Era of Empire and Globalization.* London: Earthscan, 2008.

Pollan, M. *Omnivore's Dilemma: A Natural History of Four Meals.* New York: Penguin Press, 2006.

Pollan, M. *In Defense of Food: An Eater's Manifesto*. New York: Penguin Press, 2008.
Striffler, S. *Chicken: The Dangerous Transformation of America's Favorite Food*. New Haven, CT: Yale University Press, 2005.
Virchow, D., M. Denich, A. Kuhn, and T. Beuchelt. "The Biomass-Based Value Web as a Novel Perspective on the Increasingly Complex African Agro-Food Sector." *Tropentag* September (2014): 17–19.
Weber, K. ed. *Food Inc: How Industrial Food is making us Sicker, Fatter and Poorer*. New York: Public Affairs Books, 2009.
White, Ben, S.M. Borras Jr., R. Hall, I. Scoones, and W. Wolford. "The New Enclosures: Critical Perspectives on Corporate Land Deals." *Journal of Peasant Studies* 39, nos. 3–4 (2012): 619–647.
Wittman, H., A. A. Desmarais, and N. Wiebe, eds. *Food Sovereignty: Reconnecting Food*. Nature & Community. Halifax/Cape Town: Fernwood/Pambazuka Press, 2010.
Wolford, W., S. M. Borras Jr., R. Hall, I. Scoones, and B. White. "Governing Global Land Deals: The Role of the State in the Rush for Land." *Development and Change* 44, no. 2 (2013): 189–210.
World Bank. *World Development Report 2008: Agriculture for Development*. Washington, DC: World Bank, 2007.

Exploring the 'localisation' dimension of food sovereignty

Martha Jane Robbins

International Institute of Social Studies (ISS), The Hague, Netherlands

The 'localisation' narrative is at the heart of food sovereignty in theory and practice, in reaction to the 'distance' dimension in the dominant industrial food system. But while it is a central element in food sovereignty, it is under-theorised and largely unproblematised. Using the theoretical concepts of food regime analysis, uneven geographical development and metabolic rift, the author presents an exploratory discussion on the localisation dimension of food sovereignty, arguing that not all local food systems are a manifestation of food sovereignty nor do they all help build the alternative model that food sovereignty proposes. The paper differentiates local food systems by examining character, method and scale and illustrates how local food systems rarely meet the ideal type of either food sovereignty or the capitalist industrial model. In order to address five forms of distance inherent in the global industrial food system, localisation is a necessary but not sufficient condition for food sovereignty. A more comprehensive food sovereignty needs to be constructed and may still be constrained by the context of capitalism and mediated by the social movements whence it comes.

The global industrial food system and its critics

Processes of capitalist development and its logic of profit making have shaped agriculture over time and have had a major influence on the structure and dynamics of the dominant global food system, most recently through increased trade liberalisation, corporate concentration and new technologies such as genetically modified organisms (GMOs).[1] This way of organising the food system has many long-term social implications, such as displacement and dispossession, dietary changes and a widening gap between producers and consumers; and a large impact on the environment in terms of biodiversity loss, soil depletion, deforestation and greenhouse gas emissions. An emphasis on (re)localising food production and consumption is a key component of resistance to the current industrialised and globalised structures of the food system that are at the root of these trends.

Localisation can be viewed as a direct counter to specific forms of distance in the food system. As an integral part of an alternative food system, localisation can also be viewed in opposition to the wider industrial agriculture model. Yet it is unclear if it can necessarily be equated with a democratised food system or whether all attempts at localisation can be viewed as direct critiques of the industrial model. While localisation efforts may be able to address the most accessible conception of distancing,[2] that is, physical distance between place of production and place of consumption, can they also adequately address more complex notions of distancing? Clapp uses the concept to explore both physical and abstract spaces exacerbated by the financialisation of the global food system, and it is these abstract notions of distancing that may present more difficulty for localisation efforts.[3] The ambiguous nature of defining 'local' and 'local food systems' makes this task more challenging, as these are not only defined in geographic terms but also by 'social and supply chain characteristics' or, as Feagan states, by their aspiration for '*re*spatializing and *re*configuring agricultural systems'.[4]

Peasants, social movements and civil society organisations are also putting forward alternatives such as food sovereignty, which was first presented on the world stage at the World Food Summit in 1996 by the international peasant and small-scale farmers' movement, La Via Campesina (LVC). Food sovereignty is an articulation of a radical reimagining of the food system and, as it evolves, its definition has also evolved.[5] Broadly it is the 'right of nations and peoples to control their own food systems'.[6] Food sovereignty incorporates the notion of localisation as an essential part of building alternative food systems. The declaration from the Nyéléni 2007 Forum for Food Sovereignty says that 'Food sovereignty prioritises local and national economies and markets and empowers peasant and family farmer-driven agriculture, artisanal-fishing, pastoralist-led grazing'.[7] Local markets, local economies and local production are all key aspects of a food sovereignty approach.

While academic exploration of food sovereignty has surged recently, the emphasis on localisation within food sovereignty discourse has not been unpacked in a systematic way and there are many tensions, contradictions and gaps that require consideration. First, is localisation actually (and necessarily) a challenge to the globalised, industrialised food system? Or is it merely a niche within the existing regime that allows affluent consumers more choice in their consumption habits? How does a localised food system deal with a reliance on export-oriented agriculture as the basis of an economy? Does localisation mean creating a parallel food system without altering the dominant one?

Second, while local food systems can demonstrably connect consumers more directly to producers, questions remain about who those consumers are and how far localisation efforts can reach. Can local food systems adequately feed those living in poverty and low-income situations, those who cannot afford to pay premium prices for local, ecologically produced food products? Can and will the working classes in both rural and urban settings participate in local food systems while the industrial food system continues to provide cheap food? If local production privileges fair prices for producers, can it at the same time provide affordable food for all consumers? Or is local food a contradiction of the goal of food for all?

Holt-Gimenez and Shattuck suggest that 'The challenge for food movements is to address the immediate problems of hunger, malnutrition, food insecurity and environmental degradation, while working steadily towards the structural changes needed for sustainable, equitable and democratic food systems'.[8] Can food sovereignty discourse, rooted largely in rural movements, integrate the urban food movements that often deal with practical issues of access to food and may not seek major food system transformation? Do localisation efforts have the ability to address both the practical, immediate issues in the industrial food system while simultaneously posing a substantial challenge and presenting a viable alternative to the dominant model? How central are local food systems to the realisation of transformative food sovereignty?

In this paper I will concentrate on the question: how does the food sovereignty framework, and in particular, its call for local food systems, address geographical and sectoral distancing in the current global industrial food system? I will begin by outlining the analytical tools used to inform the discussion. By situating local food initiatives within the wider food sovereignty framework, in the context of the transnational agrarian movements that developed and espouse it, I will then argue that not all local food systems are a manifestation of food sovereignty nor do they all help build the alternative model that food sovereignty proposes. By examining scale, method and character I will begin to differentiate local food systems and illustrate how they rarely meet the ideal type of either a food sovereignty model or a capitalist industrial model, and instead fall somewhere in between. Finally, I will present an investigation of five forms of distance that are inherent in the global industrial food system and examine how effective localisation efforts are in addressing these. I will argue that localisation alone is not enough and a more comprehensive understanding of food sovereignty needs to be present, although it may still be constrained by the context of capitalism and mediated by the movements from which it comes.

Conceptual exploration

Three main theoretical formulations are employed to inform the analysis that follows. Briefly 'food regime' analysis, originally formulated by Friedmann and McMichael,[9] provides a tool to analyse localisation and food sovereignty within a particular historical and political setting. Food regimes are defined as 'stable periodic arrangements in the production and circulation of food on a world scale, associated with various forms of hegemony in the world economy'.[10] McMichael argues that a corporate food regime exists as a third food regime following, first, a regime to fuel European industrialisation in the late 1800s and, second, one based on export of surpluses from the USA in the mid-twentieth century.[11] Each food regime is accompanied by a period of transition and struggle as the last regime falters and a new regime is consolidated. These periods of transition are spaces where the contestation between development models, modes of accumulation, types of agricultural systems and different food system actors can be observed; however, McMichael also recognises that food regimes themselves encompass tensions and contradictions at a particular histori-cal moment.[12] Food sovereignty and the social movements that advance it can therefore be positioned as part of the contested transition to a third, corporate food regime, or as a pivotal dynamic within an existing corporate food regime. In either formulation food sovereignty is situated as a challenge or 'counter-mobilisation' to the current global industrial food system.[13]

David Harvey's theory of uneven geographical development is a second useful theoretical construct.[14] A key idea for framing the present analysis is Harvey's understanding of theory as a dialectical process that 'is perpetually negotiating the relation between the particular and the universal, between the abstract and the concrete', which is a valuable way to consider food sover-eignty.[15] Food sovereignty is a political discourse, a proposition and, in some ways, an abstract description of a desired system of agricultural production, dis-tribution, consumption and social relations. In another sense food sovereignty is a grounded practice of concrete political, economic and social steps towards a specific vision for the food system and the actors involved in it. Harvey's asser-tion that 'Capitalist activity is always grounded somewhere',[16] that social pro-cesses are materially embedded, offers a way to think about where the industrial food system is actually located, as well as to describe the distances within it. Clear linkages can be made between Harvey's 'accumulation by dispossession' and, for example, geographical distancing understood as displacement of peas-ants from their land. But it is also possible to make the link between accumula-tion by dispossession and some forms of localisation, for example where food trends and labels such as 'local' or 'organic' are adopted by large corporations, paralleling Harvey's argument of appropriation of creativity.[17]

Finally, the metabolic rift, referring to Marx's idea of the separation between humans and nature characterised by the rupture of the natural nutrient cycle, is a mechanism for exploring distance in the current food system and the different ecological repercussions of various models of agriculture. Foster explains that this idea of ruptured metabolism was used by Marx 'to capture the material estrangement of human beings in capitalist society from the natural conditions of their existence'.[18] This concept has been used to illustrate division between urban and rural, producers and consumers, agricultural production and natural processes, as well as to argue that large-scale industrial agriculture and the development of distant markets have aggravated this rift.[19] Moore, however, challenges the use of metabolic rift as a simple binary that views capitalist development as the cause of environmental degradation and argues instead that capitalism and nature act upon each other,[20] and furthermore that capitalism develops 'through nature–society relations'.[21] Ecological damage is not a side-effect of capitalist accumulation; rather it is an intrinsic part of it. As part of a complex argument, Moore makes the point that there is a commodification tip-ping point where capitalist transformation has taken place and after which, 'nei-ther governing structures nor production systems nor the (newly transformed) forests, fields, households, and other ecologies can reproduce themselves *except through deepening participation in the circuits of capital on a world-scale*'.[22] This idea that capitalism recreates itself through its internal, yet contradictory, logics of accumulation and capitalisation poses a significant challenge to the realisation of alternative food systems.

These three distinct but interrelated clusters of theoretical lenses, namely, food regimes, uneven geographical development and metabolic rift, will inform the analysis of the specific food sovereignty problematique that I am going to tackle in this paper.

Situating food sovereignty and local food systems

Within the early framing of food sovereignty local food systems are not explicitly discussed, although the core notions of defining food and agriculture policy at a national or local level, asserting the right of peasants to exist and produce food, rejecting the industrialisation of agriculture and prioritising domestic markets suggest a strategy of localising food systems. By 2007 both the definition of food sovereignty in the Nyéléni final declaration and the six key pillars of food sovereignty outlined in the synthesis report refer to food systems shaped by local producers, the need to focus on local markets and the development of local food strategies as crucial pieces of realising food sovereignty.[23] Building and sustaining local food systems is a significant component in the food sovereignty approach yet, in their formal documents and positions, the movements for food sovereignty have not specified their vision of local food systems beyond general statements and, as Patel suggests, the definitions of food sovereignty itself are diverse, contradictory and in motion.[24]

The food sovereignty vision of local food systems remains relatively unarticulated partly because of the way in which transnational agrarian movements (TAMs) like LVC are formed and operate. Borras et al attribute the generalising tendency of TAMs to the complexity of representation. They argue that movement leaders inevitably simplify issues and stances 'to make complex realities legible [...] and manageable'.[25] Claims of representation are necessary for movements to have weight behind their proposals and demands but these claims are far from straightforward. Most TAMs are built on partial rather than full representation of a specific constituency and the degree of actual representation corresponds to the credibility and strength of the organisation or movement. In addition, representation is continuously shifting and movements go through cycles where they are more or less equipped to represent their constituencies. Finally, by claiming to represent a particular group, all those who do not identify with the group are left out and the movement does not articulate their claims.[26] Borras contends that LVC is both an actor on the international stage (and national and subnational stages via its member organisations) and at the same time an 'arena of action', a contested, dynamic internal space where positions and actions are debated, tested and negotiated, which has implications for how issues are framed and demands are made.[27]

Although the articulation of food sovereignty and local food systems by LVC and others necessarily overlooks many intricate details and leaves vital questions unresolved, broadly their expression of the food sovereignty framework views local food systems as ideally embedded in small-scale, peasant production using agroecological methods.[28] Food sovereignty seeks to move control over food systems to the local level and by its very nature, then, each local food system will demonstrate different characteristics and privilege some dimensions over others, resulting in a diversity of local food systems.[29]

Local food systems are dynamic and evolve as they develop and in this way may demonstrate changing characteristics over time, reflecting Harvey's portrayal of dialectics as constant movement between the universal and the particular, the theory and the practice.[30]

However, it is also important to note that isolating the local food system aspect of food sovereignty from its whole weakens the challenge to the corporate food regime that food sovereignty represents. Wittman et al contend that food sovereignty goes well beyond rearranging global food relations, although that is at its core, to provide an alternative model of rural development.[31] Giving credence to the possibility that food sovereignty exposes 'different concepts of modernity', Patel argues that, fundamentally, food sovereignty attempts to address deep-seated, long-standing power inequalities to achieve a 'radical egalitarianism'.[32] Patel's theorising of food sovereignty demonstrates the breadth of what is required to achieve it – including the transformation of power structures and social relations towards the elimination of 'sexism, patriarchy, racism, and class power'.[33] Reducing food sovereignty to local food systems mirrors the reductionist tendencies of the global industrial food system and therefore recreates the principles it seeks to resist. If the elements of food sovereignty are compartmentalised and dealt with separately, the transformative potential of the framework is compromised and there is a risk that the theoretical breadth of food sovereignty is lost in the concrete practice of one element. Local food systems are not sufficient on their own to challenge the global industrial food system, and a local food system – even one meeting the ideal food sovereignty type – does not constitute food sovereignty in and of itself.

Yet transformation of social relations, the food system and rural development trajectories are not possible without being grounded in the practice of food sovereignty. Is any work that is moving food system dynamics toward local or alternative models usefully contributing to food sovereignty's vision of challenging and changing the capitalist industrial model, even if only to prepare the ground? Can local food initiatives that are not controlled by farmers, that use genetically modified seeds or chemical inputs or are based on large monocultures, or that cater only to high-end consumers still be considered food sovereignty? In an attempt to tease out what constitutes local food systems based in food sovereignty and to explore the nuances, tensions and paradoxes that have yet to be solved in the articulation of food sovereignty, local food systems need to be differentiated.

Local, differentiated

The strength of local food systems is that they are based in the particular and that they can be defined in a variety of ways that appropriately suit each context. Not all forms of local are synonymous and the ambiguity of definition means that, to determine the extent to which local food systems are capable of countering distance, it is necessary to analyse their characteristics and broadly classify them along a range where local food systems within the food sovereignty framework occupy one end and local food systems within the industrial capitalist framework the other. It is important to note that the individual treatment of character, method and scale in the sections to follow is an artificial one for analytical purposes. In practice, each of these elements overlaps.

Character

Capitalism's defining features of commodity production – goods and services produced 'for market exchange in order to make a profit'[34] – and endless accumulation have left a definitive mark on agricultural production worldwide. Whether or not peasant agriculture can or does operate completely outside this logic is an unresolved debate.[35] For the purposes of classifying local food systems along a spectrum between industrial agriculture and food sovereignty, I argue that peasant character is substantively different from capitalist character. The tension between agrarianism and industrialism may be at the heart of the character differentiation between these two veins. 'The fundamental difference between industrialism and agrarianism is this: whereas industrialism is a way of thought based on monetary capital and technology, agrarianism is a way of thought based on land', writes Berry.[36] Bernstein, while arguing against this delineation, outlines the character of peasant production as described by the food sovereignty frame as 'a radically different episteme to that centred in market relations and dynamics [...] based in an ecologically wise and socially just rationality'.[37] The idea that peasants can and do operate outside the capitalist logic of accumulation yet still engage in markets and the production of a marketable surplus beyond basic subsistence production allows for an exploration of whether and how far contemporary local food system strategies act as an antithesis to agriculture organised via capitalist relations.

Accumulation by appropriation also confirms that the character of local food production and distribution is an important element of differentiation, since labels alone do not distinguish real alternatives from niches subsumed in the corporate food regime. Harvey argues that, through the process of commodification, the drive to accumulate surpluses results in appropriation of material objects and abstract ideas ('creativity' for instance) that are not generated by capital. He writes that accumulation logic 'creates a premium on the commodification of phenomena that are in other respects unique, authentic and therefore non-replicable'.[38] In many ways this explanation provides insights into the multiple manifestations of local food systems which, because of their locality, are at first glance 'unique and non-replicable'. Dominant players in the industrial food system capture alternative food system concepts quickly, however, eg Walmart offering organic foods, and turn them into marketable labels.[39] Campbell cautions that these initiatives, despite being connected to the corporate food regime, should not be entirely dismissed, as they represent 'a small but important set of counter-logics'.[40] Nevertheless, this creates obscurity in situating local food initiatives, as large corporate players appropriate labels created by alternative food movements to distinguish themselves from the mainstream food system, and therefore elevates the importance of character.

Method

The bluntest differentiation of production methods is between conventional production, which uses chemicals and synthetic fertilisers and is therefore directly connected to the corporate food regime and industrial agriculture through inputs and technology, and what can be termed 'traditional production', which includes methods from organic to agroecological production. Conventional production methods use a range of technical and synthetic interventions to increase productivity and control weeds and pests and, though widely adopted on large-scale farms in the global North, they are not confined there. Many small-scale peasant farms in the global South also use conventional methods following the introduction of Green Revolution technologies. Organisations like Alliance for a Green Revolution in Africa (AGRA), for instance, sponsor programmes to distribute hybrid seeds to small-scale farmers. Conventional methods rely on the practice of mono-cropping to ensure uniformity and durability and to exploit mechanisation.[41]

As organic food becomes more desirable, particularly for upper class consumers in the global North, parts of the organic sector, initially based on the idea of small-scale, locally based, ecologically friendly agriculture, now resemble the conventional commercial sector, including consolidated distribution networks, large-scale mono-crop production, standardisation and a heavy reliance on fossil fuels.[42] Altieri and Nicholls note that organic certification that does not address size, or what they term 'social standards', has resulted in similar arrangements and working conditions on organic farms as on conventional farms, blurring the lines between industrial and peasant agricultural systems in this respect.[43] Local food initiatives established with agroecological production methods fall more fully within the food sovereignty framework. Agroecology is based on enhancing small-scale farm productivity while conserving ecological resources through engagement in deeply rooted traditional practices and scientific knowledge of ecological processes.[44] Rosset et al summarise agroecology as a set of principles that include soil conservation and soil building, recycling of nutrients, poly-cropping and biodiversity preservation, and the use of biological mechanisms for pest control.[45] While agroecology practice is spreading through food sovereignty networks, questions remain about whether enough food can be produced, at affordable prices, to feed everyone.[46]

Scale

While it is often assumed that food system localisation implies small-scale agriculture, differentiating between different scales of production and distribution assists in distinguishing broad types of local food initiatives. Neumann observes that often the conceptualisation of scale is itself ambiguous, alternating between referring to size, level, network or site. He advocates choosing one consistent meaning of scale, but both scale as size and scale as level are useful here.[47] Iles and Montenegro's recent work adds another dimension of scale – 'relational scale', defined as 'networks of elements and processes in a complex adaptive system'.[48]

Scale as size is often invoked in relation to food production. Neumann uses the example of large-scale as interchangeably meaning 'capital intensive, spatially extensive, or national'.[49] For example, small-scale agriculture is frequently defined in opposition to large-scale agriculture based on the amount of land cultivated, yet Bernstein contends that capital intensiveness is a more useful descriptor, since farm area does not indicate the number of farm labourers needed or the capital required to start and maintain the operation.[50] While acknowledging the imprecision of the classification, industrial agriculture features increasing farm size related to mechanisation and the use of inputs, as well as increasing capitalisation, particularly in the current corporate food regime where corporate investment into the production side of the food system is becoming more commonplace.[51] Large-scale, meaning capital-intensive agricultural operations with large land area or a high number of animals, can be (and often is) used as a synonym for industrial agriculture. In fact, van der Ploeg calls increasing scale an 'indispensable ingredient of industrialization'.[52] Linking large-scale to industrial agriculture does not preclude the sale of industrial production within a local food system, although large-scale production typified by industrial methods is likely to be connected into larger distribution networks and more distant markets, as is illustrated through the three food regimes and their respective global commodity flows.

Scale as level is also a useful conceptualisation. The lower down the scale or closer to the household level a food system is, the more local it is. In this version of scale, small-scale means producing for a household or a market within the community while large-scale means producing for levels further up the scalar chain, such as international markets. The further down the local scale a food system is, the shorter the supply chain.[53] Using this framing, a farmers market where farmers are directly selling to customers is more local than a large grocery chain offering regional produce. The lower down the scale local food systems are, the more able they should be to bridge the metabolic rift, both literally and metaphorically. This is not always straightforward in practice, however, particularly as efforts to build local food systems grapple with questions of how to scale up these initiatives to include more consumers and larger geographical areas without losing focus on maintaining the embeddedness of these food systems. Friedmann relates the case of public procurement of local food in Toronto as a mechanism for shortening supply chains by utilising a third party certifying body as the link between private caterers at a university and local farmers. She contends that this is a successful case of increasing the scale of local food systems. Yet, in the process of scaling up, the direct linkage between producer and consumer is unavoidably a few steps removed.[54] While scaling up localisation presents difficulties for maintaining its benefits, local food systems need to reach a large number of consumers in order to challenge the industrial food system.

In both these cases it is possible to generalise that large-scale more readily corresponds to an industrial, capitalist framework and small-scale to a food sovereignty framework. However, as Iles and Montenegro argue, 'One cannot adopt a fixed, small-scale approach to confront such a flexible, 'many-headed beast' as capitalist agriculture',[55] which is demonstrated by the complexity of attempting to achieve global, system-wide change through small-scale agriculture.

Figure 1 presents a three-dimensional way to visualise the complexity of differentiating local food systems.[56] The food sovereignty framework ideal type is small-scale, agroecological, peasant production. However, most local food system initiatives have characteristics that are situated somewhere along the three axes of scale, character and method. Farmers markets may rely on relatively small-scale peasant production from farms using nitrogen fertilisers or other inputs, thereby fitting somewhere in the 'small-scale, peasant, conventional' box on the diagram. Community-Supported Agriculture (CSA) initiatives may use agroecological methods on small-scale farms but hire wage labourers in the busy

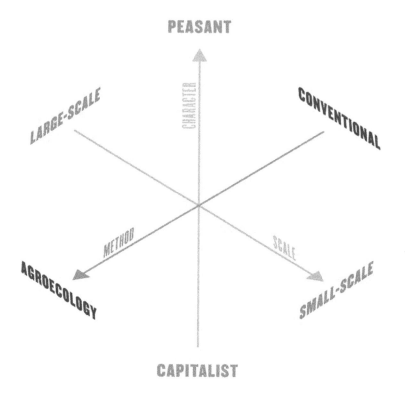

Figure 1.   Differentiating local food systems.
Source: Author's own diagram; graphic design: Aylwin Lo.

season, and therefore could be placed in the 'small-scale, agroecological, capitalist' box. Public procurement initiatives may require certified organic food but not specify the scale of production, etcetera. Attempts to differentiate various forms of local can benefit from these explorations of scale, method and character, although they do not lead to a convenient and simple binary between the two models. Rather, what emerges is a gradient where, in practice, local food initiatives may fall closer to one model than the other, may demonstrate seemingly contradictory characteristics, or may move between the categories over time. In this way, localisation is not automatically synonymous with food sovereignty.

Food sovereignty deals with distance through localisation

How does food sovereignty, and in particular its focus on local food systems, address the geographical and sectoral distances in the global industrial food system? In the context of a corporate food regime arranged around market principles arbitrated by international institutions like the World Trade Organization (WTO), and favouring industrial agriculture and corporate supply chains, food sovereignty appears as both a contradiction and a proposal for reorienting food and agriculture. The location of food sovereignty as a counter-movement to the corporate food regime and its particular logic of capital accumulation is mediated by the nature of TAMs, as illustrated above. Therefore, it is worthwhile analysing the techniques and effectiveness of food sovereignty movements and actors and their attendant local food system strategies to counteract the distinct forms of distance outlined below, as the ability of food sovereignty initiatives to respond to the global industrial food system in a concrete, systematic way is uncertain. Moore's claim of a commodification tipping point beyond which production systems can only reproduce themselves by becoming further entrenched in capitalist processes looms large.[57]

Production and consumption

Perhaps the most obvious form of distance in the global industrial food system is the geographic or physical distance between where food is produced and where it is consumed. By looking at how food regimes have reshaped the relations between commodities, production methods and wider processes of capitalist development, it becomes clear that the industrial food system perpetuates distance between production and consumption and between producers and consumers. Distance between producers and consumers is created when consumers are detached from those who produce their food and from the processes involved in production and processing, as the metabolic rift suggests. This divide is produced, in part, by the commodification of food, which 'creates an abstract and disembodied notion of food' through standardisation, processing and global distribution.[58] Transforming food into a commodity is necessary for the functioning of the industrial food system, as disembedding food from its social, cultural, geographic and ecological aspects allows food to be vastly exchangeable and severely altered from its original state.[59] Commodification of food and the processes that have enabled it conceal the producers (and the production processes, maintains Campbell)[60] from consumers. Friedmann argues that 'the dominant tendency is towards distance and durability, the suppression of particularities of time and place in both agriculture and diets. More rapidly and deeply than before, transnational agrifood capitals disconnect production from consumption and relink them through buying and selling'.[61]

Locality and seasonality are cited as responses to features of geographical distancing in the industrial food system and, in this way, localisation is proposed as an oppositional force to distance and durability.[62] Local food movements tackle the distance between producers and consumers through direct marketing and the social re-embedding of food systems. Hinrichs asks difficult questions about who benefits in these local food transactions, however, and warns that social embeddedness is often taken for granted and generalised, when it should be assessed more vigorously.[63] Local food movements may heighten distancing through their 'missionary impulses', according to Guthman's scathing critique of local food system practices in the USA as racially marginalising and exclusionary.[64] Allen and Wilson suggest that many local food initiatives may reinforce existing inequalities of race and class rather than challenging them, because of the focus on consumer choice that ignores historically constructed inequalities. They posit that food sovereignty and other initiatives centred in the global South are more attuned to the importance of dealing with inequality.[65] In contrast, Holt-Gimenez and Shattuck characterise the food justice movements as more concerned with marginalised people and their access to food and question the notion that they are predominantly made up of elites. They portray the food sovereignty movement as more concerned with structural transformation and redistribution of productive resources, often with a class lens, although this is not always clear.[66] Local food systems based in food sovereignty reduce the distance between production and consumption but this holds true for virtually any type of local food system. The question that remains is how much consumption, of what kinds of food, and by whom can be captured by local food systems? In many ways food sovereignty is more narrowly focused on the outcomes for producers than consumers, although the framework actively promotes the right to food (connected to its de-commodification) and advocates space for marginalised communities to reclaim their own food systems.

Distant markets

By distancing production from consumption it is inevitable that agricultural products will often be sold in a place far away from where they were grown, necessitating a chain of intermediaries and market relations with 'innumerable points for the extraction of value and surplus value'.[67] The globalisation of markets also translates into an abstraction whereby the market becomes an entity in itself rather than a space of transaction governed by the buyer and the seller. 'Financialization promotes a new kind of distancing by encouraging a greater abstraction of agricultural commodities from their physical form', states Clapp.[68] Edelman contends that the move from market, understood as a specific place, to market, understood as a metaphor or abstract space, required distancing the economy from society and hiding the actors and institutions that shape economic transactions.[69] Following Harvey's conception of the material embeddedness of social processes, the intangible global agricultural commodities market, though complex, is still dealing in material products that have been grown in a particular place and will be consumed in a particular place.[70] The consequences of this abstract market are also embedded in the material world; however, increasing financialisation of agriculture has rendered cause and effect more opaque and therefore more challenging to resist.[71]

The food sovereignty framework does not preclude trade, although it promotes the removal of corporate control, free trade agreements and the flooding of domestic markets with below-cost of production food from global circuits of trade, and calls for agriculture to be taken out of the purview of the WTO.[72] The challenge for its proponents is that this dominant global paradigm still exists and implementation of food sovereignty occurs within the current global context even as it seeks to change it. Food sovereignty counters the abstraction of markets by pursuing primarily domestic and local markets and by seeking to ground market relations, put food producers at the centre and nutritional needs ahead of the market.[73] Local, direct marketing strategies do not automatically encompass these values, however. For example, farmers' markets have been critiqued by Hinrichs as recreating commodity relations, despite higher levels of social embeddedness than with conventional markets. Food box programmes and CSAs (where customers buy in to the risks of the farm at the beginning of the growing season and in return receive fresh produce throughout the harvest)[74] are more likely to break down the commodity relations of conventional markets and create 'an alternative *to the* market'[75], but this largely Northern model reaches very few consumers. Burnett and Murphy argue that the food sovereignty position on trade is too vague, given that it is still vital for peasant livelihoods and urban food security and that it could be shaped by food sovereignty discourse.[76] This raises the question: to what extent can local, direct marketing strategies eschew the imperatives of capitalist market relations and create an alternative set of concrete market relationships articulating food sovereignty?

Peasants from their land

The dispossession of peasants from their land and the migration and displacement that results constitutes a third way of viewing geographical distance. Araghi argues that enclosure, privatisation and dispossession are defining features of capitalism, not only in its infancy but also as a continuous operational logic.[77] While it is important to note that it does not occur at every site of accumulation, Harvey states that dispossession is 'a necessary condition for capitalism's survival'.[78] Dispossession not only occurs to those caught struggling against the imposition of cheap imports flooding their national and local markets, it also happens to those who engage and participate in the same industrial food system that is eventually responsible for their dispossession. The dispossessed may be 'adversely incorporated' into the global food system, where they are marginalised and exploited,[79] for example as migrant labourers on highly industrialised, single commodity-driven farms or as contract farmers integrated into corporate production systems. Participation is often forced by the search for higher yields to offset low prices or the consolidation of processers who prefer to contract production on their terms or by the difficulty of unhooking from the industrial system once you are connected to it through inputs and other means. In Canada farm debt increased two and a half times between 1988 and 2007 and more than 10% of farms disappeared between 2006 and 2011, despite Canadian farmers engaging wholeheartedly in the industrial food system – Canadian agri-food exports tripled between 1988, when Canada signed the first free trade agreement with the USA, and 2007.[80]

To increase resilience against processes of dispossession, food sovereignty claims local control over food systems. The process of localising food systems has a more tenuous connection to peasants who are already dispossessed, however. Dispossession and the resulting displacement or adverse incorporation weakens peasant organisations, reducing their representative authority and their ability to control local food systems provisioning nearby urban centres. This is illustrated by Edelman's analysis of the rise and fall of Central America's transnational peasant alliance, which struggled to maintain its unity and strength in a context of deteriorating opportunities for small-scale peasants, increasing out-migration and other significant changes in the countryside. 'Migration frequently undermines the capacity for political action', he contends.[81] Borras et al reference a 'persistent and troubling divide' between escalating migration and the lack of representation and focus on migrant labourers by LVC and other TAMs.[82] While LVC does claim migration and rural workers as one of its core issue clusters, some of the difficulty may be attributed to the different class positions of small-scale farmers and farmworkers, as Bernstein asserts.[83]

Rural and urban

Definitions of rural and urban may be fluid, as rural people migrate to cities to find work and urban people move into rural areas in search of affordable housing or different lifestyles. Rigid definitions also negate the peri-urban and small town spaces between farm and metropolis and the flows of goods, services and people that exist between them.[84] Yet rural communities dependent on agriculture face unique challenges stemming from economic uncertainty, out-migration, urbanisation and the loss or underdevelopment of vital services. Rural and urban distancing is linked to historical and contemporary arguments about prioritising industry (primarily urban) over agriculture (rural), or vice versa, as a development strategy. Kay suggests that, to bridge the rural–urban divide, development strategies need to focus on synergies and interactions between the two sectors rather than extracting surplus from one to fuel the other.[85]

Bernstein asks perhaps the most critical question of food sovereignty and local food systems related to shrinking the distance between rural and urban: 'how plausible are the claims of agrarian "counter-movements" and their champions that a return to "low-input" small-scale family farming [...] can feed a world population so many times larger, and so much more urban, than the time when "peasants" were the principal producers of the world's food?'[86] According to the United Nations, the majority of the world's population now lives in urban areas and the number is expected to climb 2.6 billion people by 2050, which represents the entire increase in world population until then plus the migration of 0.3 billion rural inhabitants to urban areas.[87] In this context the question of how local food systems will feed cities becomes especially relevant. Food sovereignty proponents argue that peasant agricultural systems still produce the majority of the world's food (although this is contested) and that agroecological peasant production is more productive and more capable of adapting to and mitigating climate change than the industrial model. Therefore, food systems built around food sovereignty are more capable of dealing with both hunger and urbanisation.[88] Nevertheless, the question of feeding large urban centres has not been dealt with systemically. Rosset et al's Cuban case study illustrates that scaling up the farmer-to-farmer agroecology methodology still poses significant challenges, particularly in the face of structural barriers.[89] Urban agriculture, a facet of local food systems that has been largely unexplored by the food sovereignty movement, has recently gained attention as part of the answer to bridging this gap. The September 2012 Nyéléni Newsletter stated that food sovereignty had an urban dimension and listed urban and peri-urban agriculture as a food sovereignty strategy.[90]

Agriculture and nature

The separation of agriculture from natural processes is the basis of the metabolic rift, described above.[91] In the first food regime agricultural industrialisation was characterised by commercial commodity production, increased mechanisation, privatisation of land and the expansion of 'ecological exploitation' outside Europe.[92] During the second food regime synthetic chemicals and fertilisers were incorporated to increase production and to counter the effects of nutrient-deficient soils. The current food regime follows the same logic of expansion, industrialisation and separation of food systems from their ecological base. Green Revolution technologies are intensely promoted in agricultural systems in the global South. The use of fossil fuels, new technologies such as genetically modified seeds, and the intensification of livestock production – linked to what Weis terms 'an expanding ecological hoofprint',[93] constitute a spiral that reinforces itself, requiring more and more investment in industrial methods to maintain production increases and perpetual accumulation. According to van der Ploeg, 'instead of being built on ecological capital, farming has become dependent upon industrial and financial capital'.[94] Agriculture has come loose from its ecological foundations in a process of constant and chronic distancing.

Harvey's explanation of how capital accumulation innately seeks to speed up production, distribution and consumption to reduce the costs associated with distance is relevant to Moore's position that capitalism develops through the relations between nature and society.[95] Moore provides a list of examples, such as accelerating the time it takes to grow chickens and speeding up milk production in cows through the use of hormones.[96] This endeavour to overcome space and time is reliant on the type of capital that itself is delinked from materiality. While the limitations of distance are tackled in this logic, in another sense the distance between agriculture and natural systems is aggravated. By accepting Moore's framing of nature–society relations and the premise that nature is also acting on capitalism, it is still possible to assert that the logic of capital accumulation within this relationship has fundamentally altered the link between agriculture and nature.

While local food systems based in food sovereignty are not the only local food efforts that take ecological questions into consideration, food sovereignty proposes a major shift in nature–society relations and a step toward mitigating the metabolic rift. McMichael maintains that food sovereignty challenges the ecological and social impacts of the industrial model and 'engages modern science and technology' in new ways.[97] According to Schneider and McMichael, metabolic rift theorising often excludes taking into account the knowledges of local places and agricultural methods that were and are lost as the rift widens.[98] In contrast, Altieri and Nicholls offer a number of examples of contemporary agroecology in Latin America integrating ancient techniques, demonstrating that agroecological methods are intrinsically localised to each specific region and natural environment; they seek to bridge the metabolic rift understood as a break with ecosystems and with the reproduction of knowledges.[99] The food sovereignty model based on agroecological production methods has a different relationship to nature than does its industrial counterpart promoting biodiversity and seed-saving, low emission agriculture and shortened supply chains. LVC's most active, grounded local food system work is through agroecology training.

Conclusion

This paper represents a preliminary attempt to dissect the role of local food systems within the food sovereignty framework and analyse how far they go in challenging the current model of industrial agriculture. Although the initial findings illustrate that not all local food systems are manifestations of food sovereignty or even operate as alternatives to the corporate food regime, neither can they be unequivocally categorised as part of the capitalist industrial model. In practice most local food initiatives fall somewhere in between the two arche-types. Food sovereignty provides an alternative, both politically and practically, but its demands are mediated both through the nature of the capitalist system and corporate food regime in which it exists and by the dynamics intrinsic to transnational agrarian movements that have to reconcile diverse positions and the voices of many actors. These factors mean that, while food sovereignty moves towards addressing the five identified forms of distancing that character-ise the global industrial food system, some questions remain unresolved by the current theory and practice of food sovereignty.

While much of the local food system research for this paper was focused on the global North, particularly North America, more research on local food sys-tems in the global South, including specific case studies as well as larger system analyses, is needed. Other dimensions of distance (including complex issues of scale), the role of the state and other institutions as well as work that considers these questions more explicitly within current development discourses are all avenues of future research.

Although local food initiatives do not fully address the distances elaborated above, the discourse and localising practises of food sovereignty constitute a major contribution to uncovering and challenging the complex nexus of issues inherent in the current food regime. Food sovereignty represents a paradigm shift in rural development thinking and, through the TAMs that advocate it, has rightly claimed a space on the international stage and become an important part of the discourse on restructuring the food system.

Notes

1. Bernstein, *Class Dynamics*, 79, 82–83; Holt-Gimenez and Shattuck, "Food Crisis," 111; and Borras, "Agrarian Change," 6–9.
2. The concept of distancing has roots in Kneen, *From Land to Mouth*, 27, who defines it as a separation between human nutrition and food production via industrial processes. Princen, "Distancing," 116, some-what similarly defines it as 'the separation between primary resource extraction decisions and ultimate consumption decisions'.
3. Clapp, "Financialization," 798.
4. Martinez et al., *Local Food Systems*, 3; and Feagan, "The Place of Food," 24 (emphasis in the original).
5. Desmarais, *La Via Campesina*, 34.
6. Wittman et al., "The Origins and Potential," 2.
7. "Declaration of Nyéléni."
8. Holt-Gimenez and Shattuck, "Food Crisis," 132.
9. Friedmann and McMichael "Agriculture and the State System," 93–117.
10. McMichael, "A Food Regime Analysis," 281.
11. McMichael, "A Food Regime Genealogy," 141–148.
12. Ibid., 146–147.
13. Ibid., 148; and Holt-Gimenez and Shattuck, "Food Crisis," 109–144.
14. The concept of uneven development was first posited by Neil Smith, *Uneven Development*, in 1984.
15. Harvey, *Spaces of Global Capitalism*, 76.

16. Ibid, 78.
17. Ibid, 92.
18. Foster, "Marx's Theory," 383.
19. Ibid; Clark and Foster, "The Dialectic," 127; and Wittman, "Reworking the Metabolic Rift," 808.
20. This position is not new. It appears in Harvey, *Spaces of Global Capitalism*, and earlier in Smith, *Uneven Development*.
21. Moore, "Transcending the Metabolic Rift," 2.
22. Ibid., 32 (emphasis in the original).
23. "Declaration of Nyéléni"; and *Nyéléni 2007 Forum for Food Sovereignty Synthesis Report*.
24. Patel, "Grassroots Voices."
25. Borras et al., "Transnational Agrarian Movements," 186.
26. Ibid., 182–186.
27. Borras, "The Politics of Agrarian Movements," 779, 783.
28. In the past decade LVC has been placing more emphasis on a wider variety of food producers, such as fisherfolk and pastoralists, in addition to peasants. *Nyéléni 2007 Food Forum for Food Sovereignty Synthesis Report*.
29. Ibid., 1.
30. Harvey, *Spaces of Global Capitalism*, 76.
31. Wittman et al., "The Origins," 4–5.
32. Desmarais, *La Via Campesina*, 39; and Patel, "Grassroots Voices," 670–671.
33. Patel, "Grassroots Voices," 670–671.
34. Bernstein, *Class Dynamics*, 25.
35. See Bernstein, "Food Sovereignty," for a contemporary version of this debate.
36. Berry, "The Whole Horse," 42.
37. Bernstein, "Food Sovereignty," 1041.
38. Harvey, *Spaces of Global Capitalism*, 92.
39. Campbell, "Breaking New Ground," 316.
40. Ibid., 318.
41. Friedmann, "Distance and Durability."
42. Guthman, "Raising Organic."
43. Altieri and Nicholls, "Agroecology," 34–36.
44. Altieri and Nicholls, "Scaling up Agroecological Approaches," 476.
45. Rosset et al., "The Campesino-to-Campesino Agroecology Movement," 163.
46. Bernstein, "Food Sovereignty," 1051–1053. Jansen, "The Debate on Food Sovereignty," provides a useful discussion on the productivity debate, particularly the claim that agro-ecology can be productive enough to feed the world.
47. Neumann, "Political Ecology," 405–406.
48. Iles and Montenegro, "Sovereignty at what Scale?," 2.
49. Neumann, "Political Ecology," 404.
50. Bernstein, *Class Dynamics*, 93.
51. Van der Ploeg, "The Food Crisis," 100.
52. Ibid.
53. Scaling up or insertion into chains can also be used to mean more integration of peasants into global markets and input and technology circuits, as is evidenced by the World Bank, *World Development Report 2008*.
54. Friedmann, "Scaling Up," 389.
55. Iles and Montenegro, "Sovereignty at what Scale?," 7.
56. Imagine the diagram as a set of boxes.
57. Moore, "Transcending the Metabolic Rift," 32.
58. Jacobsen, "The Rhetoric of Food," 67.
59. Ibid; and Campbell, "Breaking New Ground," 310.
60. Campbell, "Breaking New Ground," 311.
61. Friedmann, "Distance and Durability," 379.
62. Ibid., 380.
63. Hinrichs, "Embeddedness," 297; and Hinrichs, "The Practice," 36.
64. Guthman, "'If They only Knew'," 395.
65. Allen and Wilson, "Agrifood Inequalities," 537.
66. Holt-Gimenez and Shattuck, "Food Crisis," 115, 131–132.
67. Harvey, *Spaces of Global Capitalism*, 97.
68. Clapp, "Financialisation," 800.
69. Edelman, "Bringing the Moral Economy Back In," 332.
70. Harvey, *Spaces of Global Capitalism*, 78; and Iles and Montenegro, "Sovereignty at what Scale?," 7.
71. Clapp, "Financialization," 801.
72. "Declaration of Nyéléni."
73. Allen and Wilson, "Agrifood Inequalities," 537.

74. Feagan, "Direct Marketing," 161–162.
75. Hinrichs, "Embeddedness," 295 (emphasis in the original).
76. Burnett and Murphy, "What Place for International Trade?"
77. Araghi, "The Invisible Hand," 120.
78. Harvey, *Spaces of Global Capitalism*, 91.
79. Du Toit, "'Social Exclusion'," 1002–1004.
80. Statistics Canada, *2011 Farm*; and NFU, "'Free Trade.'"
81. Edelman, "Transnational Organizing," 249.
82. Borras et al., "Transnational Agrarian Movements," 185.
83. LVC website, accessed May 31, 2014. http://viacampesina.org/en/; and Bernstein, *Class Dynamics*.
84. Von Braun, "Rural–Urban Linkages," 1–3; and Kay, "Development Strategies," 122.
85. Kay, "Development Strategies," 115.
86. Bernstein, *Class Dynamics*, 122–123.
87. DESA, *World Urbanization Prospects*, 1.
88. LVC, *Sustainable Peasant and Family Farm Agriculture*; ETC Group, "'Who Will Feed Us?'"; and LVC, *Small Scale Sustainable Farmers*.
89. Rosset et al., "The Campesino-to-Campesino Agroecology Movement," 185.
90. "Food and Cities."
91. Foster, "Marx's Theory"; Moore, "Transcending the Metabolic Rift"; Clark and Foster, "Ecological Imperialism"; and Wittman, "Reworking the Metabolic Rift."
92. Clark and Foster, "Ecological Imperialism," 312; and Friedmann and McMichael, "Agriculture and the State System," 101.
93. Weis, "The Accelerating Biophysical Contradictions," 317.
94. Van der Ploeg, "The Food Crisis," 100.
95. Harvey, *Spaces of Global Capitalism*, 100; and Moore, "Transcending the Metabolic Rift."
96. Moore, "Transcending the Metabolic Rift," 14.
97. McMichael, "Peasants make their own History," 505.
98. Schneider and McMichael, "Deepening," 477.
99. Altieri and Nicholls, "Scaling up Agroecological Approaches," 476.

Bibliography

Allen, P., and A. B. Wilson. "Agrifood Inequalities: Globalization and Localization." *Development* 51, no. 4 (2008): 534–540.
Altieri, M. A., and C. I. Nicholls. "Agroecology: Rescuing Organic Agriculture from a Specialized Industrial Model of Production and Distribution." *Policy Matters: Trade, Environment and Investment – Cancun and Beyond*, no. 11. IUCN Commission on Environmental, Economic, & Social Policy, September 2003, 34–41.
Altieri, M. A., and C. I. Nicholls. "Scaling up Agroecological Approaches for Food Sovereignty in Latin America." *Development* 51, no. 4 (2008): 472–480.
Araghi, F. "The Invisible Hand and the Visible Foot: Peasants, Dispossession and Globalization." In *Peasants and Globalization: Political Economy, Agrarian Transformation and Development*, edited by A. H. Akram-Lodhi and C. Kay, 111–147. New York: Routledge, 2012.
Bernstein, H. *Class Dynamics of Agrarian Change*. Sterling, VA: Kumarian Press, 2010.
Bernstein, H. "Food Sovereignty via the 'Peasant Way': A Sceptical View." *Journal of Peasant Studies* 41, no. 6 (2014): 1031–1063.
Berry, W. "The Whole Horse: The Preservation of the Agrarian Mind." In *The Fatal Harvest Reader: The Tragedy of Industrial Agriculture*, edited by A. Kimbrell, 39–49. Sausalito, CA: Foundation for Deep Ecology, 2002.
Borras Jr., S. M. "Agrarian Change and Peasant Studies: Changes, Continuities and Challenges – An Introduction." *Journal of Peasant Studies* 36, no. 1 (2009): 5–31.
Borras Jr., S. M. "The Politics of Transnational Agrarian Movements." *Development and Change* 41, no. 5 (2010): 771–803.
Borras Jr., S. M., M. Edelman, and C. Kay. "Transnational Agrarian Movements: Origins and Politics, Campaigns and Impact." *Journal of Agrarian Change* 8, nos. 2–3 (2008): 169–204.
Burnett, K., and S. Murphy. "What Place for International Trade in Food Sovereignty?" *Journal of Peasant Studies* 41, no. 6 (2014): 1065–1084.
Campbell, H. "Breaking New Ground in Food Regime Theory: Corporate Environmentalism, Ecological Feedbacks and the 'Food from Somewhere' Regime?" *Agriculture and Human Values* 26, no. 4 (2009): 309–319.
Clapp, J. "Financialization, Distance and Global Food Politics." *Journal of Peasant Studies* 41, no. 5 (2014): 797–814.
Clark, B., and J. B. Foster. "The Dialectic of Social and Ecological Metabolism: Marx, Meszaros, and the Absolute Limits of Capital." *Socialism and Democracy* 24, no. 2 (2010): 124–138.

Clark, B., and J. B. Foster. "Ecological Imperialism and the Global Metabolic Rift: Unequal Exchange and the Guano/Nitrates Trade." *International Journal of Comparative Sociology* 50, nos. 3–4 (2009): 311–334.

"Declaration of Nyéléni." Selingue, Mali, 27 February 2007. www.nyeleni.org/spip.php?article290.

DESA. *World Urbanization Prospects: The 2011 Revision.* New York: Department of Economic and Social Affairs – Population Division of the United Nations, 2012.

Desmarais, A. A. *La Via Campesina.* Halifax: Fernwood Publishing, 2007.

Du Toit, A. "'Social Exclusion' Discourse and Chronic Poverty: A South African Case Study." *Development and Change* 35, no. 5 (2004): 987–1010.

Edelman, M. "Bringing the Moral Economy back in…to the Study of 21st-Century Transnational Peasant Movements." *American Anthropologist* 107, no. 3 (2005): 331–345.

Edelman, M. "Transnational Organizing in Agrarian Central America: Histories, Challenges, Prospects." *Journal of Agrarian Change* 8, nos. 2–3 (2008): 229–257.

ETC Group. "'Who Will Feed Us?' Questions for the Food and Climate Crises." *Communique* 102 (2009): 1–31.

Feagan, R. "Direct Marketing: Towards Sustainable Local Food Systems?" *Local Environment: The International Journal of Justice and Sustainability* 13, no. 3 (2008): 161–167.

Feagan, R. "The Place of Food: Mapping out the 'Local' in Local Food Systems." *Progress in Human Geography* 31, no. 1 (2007): 23–42.

"Food and Cities." *Nyeleni Newsletter*, no. 11 (2012). http://nyeleni.org/spip.php?page=NWrub.en&id_rubrique=105.

Foster, J. B. "Marx's Theory of Metabolic Rift: Classical Foundations for Environmental Sociology." *American Journal of Sociology* 105, no. 2 (1999): 366–405.

Friedmann, H. "Distance and Durability: Shaky Foundations of the World Food Economy." *Third World Quarterly* 13, no. 2 (1992): 371–383.

Friedmann, H. "Scaling Up: Bringing Pubic Institutions and Food Service Corporations into the Project for a Local, Sustainable Food System in Ontario." *Agriculture and Human Values* 24 (2007): 389–398.

Friedmann, H., and P. McMichael. "Agriculture and the State System: The Rise and Decline of National Agricultures, 1870 to the Present." *Sociologia Ruralis* 29, no. 2 (1989): 93–117.

Guthman, J. "'If They only Knew': Color Blindness and Universalism in California Alternative Food Institutions." *Professional Geographer* 60, no. 3 (2008): 387–397.

Guthman, J. "Raising Organic: An Agro-ecological Assessment of Grower Practices in California." *Agriculture and Human Values* 17, no. 3 (2000): 257–266.

Harvey, D. *Spaces of Global Capitalism: Towards a Theory of Uneven Geographical Development.* London: Verso, 2006.

Hinrichs, C. C. "Embeddedness and Local Food Systems: Notes on Two Types of Direct Agricultural Market." *Journal of Rural Studies* 16 (2000): 295–303.

Hinrichs, C. C. "The Practice and Politics of Food System Localization." *Journal of Rural Studies* 19 (2003): 33–45.

Holt-Gimenez, E., and A. Shattuck. "Food Crisis, Food Regimes and Food Movements: Rumblings of Reform or Tides of Transformation?" *Journal of Peasant Studies* 38, no. 1 (2011): 109–144.

Iles, A., and M. Montenegro de Wit. "Sovereignty at What Scale? An Inquiry into Multiple Dimensions of Food Sovereignty." *Globalizations* (published online 16 September 2014), 1–17. http://dx.doi.org/10.1080/14747731.2014.957587.

Jacobsen, E. "The Rhetoric of Food." In *The Politics of Food*, edited by M. E. Lien and B. Nerlich, 59–78. Oxford: Berg, 2004.

Jansen, K. "The Debate on Food Sovereignty Theory: Agrarian Capitalism, Dispossession and Agroecology." *Journal of Peasant Studies* 42, no. 1 (2015): 213–232.

Kay, C. "Development Strategies and Rural Development: Exploring Synergies, Eradicating Poverty." *Journal of Peasant Studies* 36, no. 1 (2009): 103–137.

Kneen, B. *From Land to Mouth: Understanding the Food System.* Toronto: NC Press, 1989.

La Via Campesina (LVC). *Small Scale Sustainable Farmers are Cooling down the Earth* Jakarta: La Via Campesina, 2009.

LVC. *Sustainable Peasant and Family Farm Agriculture can Feed the World.* Jakarta: La Via Campesina, 2010.

Martinez, S., M. Hand, M. DaPra, S. Pollack, K. Ralston, T. Smith, S. Vogel, S. Clark, L. Lohr, S. Low, and C. Newman. *Local Food Systems: Concepts, Impacts, and Issues.* ERR 97. U.S. Department of Agriculture, Economic Research Service, May 2010.

McMichael, P. "A Food Regime Analysis of the 'World Food Crisis'." *Agriculture and Human Values* 26, no. 4 (2009): 281–295.

McMichael, P. "A Food Regime Genealogy." *Journal of Peasant Studies* 36, no. 1 (2009): 139–169.

McMichael, P. "Peasants make their own History, but not just as they Please…" *Journal of Agrarian Change* 8, nos. 2–3 (2008): 205–228.

Moore, J. W. "Transcending the Metabolic Rift: A Theory of Crises in the Capitalist World-ecology." *Journal of Peasant Studies* 38, no. 1 (2011): 1–46.

Neumann, R. P. "Political Ecology: Theorizing Scale." *Progress in Human Geography* 33, no. 3 (2009): 398–406.

National Farmers Union (NFU). "'Free Trade': Is it working for Farmers? Comparing 2007 to 1988." *Union Farmer Monthly* Special Edition 57, no. 6 (2007).

Nyeleni 2007 Forum for Food Sovereignty Synthesis Report. Selingue, Mali, March 31, 2007. www.nyeleni. org/spip.php?article334.

Patel, R. "Grassroots Voices: What does Food Sovereignty look Like?" *Journal of Peasant Studies* 36, no. 3 (2009): 663–706.

Princen, T. "Distancing: Consumption and the Severing of Feedback." In *Confronting Consumption*, edited by T. Princen, M. Maniates, and K. Conca, 103–131. Cambridge, MA: MIT Press, 2002.

Rosset, P. M., B. M. Sosa, A. M Roque Jaime, and D. R. Avila Lozano. "The Campesino-to-Campesino Agroecology Movement of ANAP in Cuba: Social Process Methodology in the Construction of Sustainable Peasant Agriculture and Food Sovereignty." *Journal of Peasant Studies* 38, no. 1 (2011): 161–191.

Schneider, M., and P. McMichael. "Deepening, and Repairing, the Metabolic Rift." *Journal of Peasant Studies* 37, no. 3 (2010): 461–484.

Smith, N. *Uneven Development: Nature, Capital and the Production of Space.* Oxford: Basil Blackwell, 1984.

Statistics Canada. *2011 Farm and Farm Operator Data.* 2012. www.statcan.gc.ca/pub/95-640-x/2012002/01-eng.htm#I.

Van der Ploeg, J. D. "The Food Crisis, Industrialized Farming and the Imperial Regime." *Journal of Agrarian Change* 10, no. 1 (2010): 98–106.

Von Braun, J. "Rural–Urban Linkages for Growth, Employment, and Poverty Reduction." Keynote Speech to Plenary I at the Fifth International Conference on the Ethiopian Economy, June 7–9, 2007.

Weis, T. "The Accelerating Biophysical Contradictions of Industrial Capitalist Agriculture." *Journal of Agrarian Change* 10, no. 3 (2010): 315–341.

Wittman, H. "Reworking the Metabolic Rift: La Via Campesina, Agrarian Citizenship, and Food Sovereignty." *Journal of Peasant Studies* 36, no. 4 (2009): 805–826.

Wittman, H., A. A. Desmarais, and N. Wiebe. "The Origins and Potential of Food Sovereignty." In *Food Sovereignty: Reconnecting Food, Nature and Community*, edited by H. Wittman, A. A. Desmarais, and N. Wiebe, 1–14. Halifax: Fernwood Publishing, 2010.

World Bank. *World Development Report 2008: Agriculture for Development Overview.* Washington, DC: World Bank, 2007.

Food sovereignty, food security and fair trade: the case of an influential Nicaraguan smallholder cooperative

Christopher M. Bacon

Department of Environmental Studies and Sciences, Santa Clara University, USA

The relationships among trade, food sovereignty and food security are underexplored. I conducted qualitative research with an influential cooperative to identify lessons that food sovereignty (FS) scholars could learn from fair trade and food security, and explore linkages among these projects. First, most co-op leaders and farmers view these projects as complementary, not contradictory. Second, state-led agrarian reforms and co-ops increase access to land, markets, water, forests and pasture, which have reduced – but not eliminated – seasonal hunger. Third, these diversified fair trade coffee-exporting smallholders could be part of a FS agenda. However, the split in fair trade suggests that only specific versions of fair trade are compatible with FS. Fourth, capable cooperatives can enhance fair trade and FS goals, and food security outcomes. Fifth, organised smallholders resisting the fair trade split could learn from the FS social movement's strategies. Food insecurity remains a persistent challenge to both approaches.

Introduction

Two recent conferences analysed the rise of food sovereignty and assessed the implications of this evolving concept, practice and social movement. The definition offered at a Yale University conference holds that food sovereignty is 'the right of peoples to democratically control or determine the shape of their food systems, and to produce sufficient and healthy food in culturally appropriate and ecological ways in and near their territory'.[1] The food sovereignty social movement (FSM) agenda is expansive and changing. It prioritises land and water access, sustainable peasant agriculture, biodiversity, justice, gender equity, participatory democratic governance, and rural and indigenous peoples' collective human rights.[2] The FSM works against land-grabbing and centralised corporate

FOOD SOVEREIGNTY

and state control of the dominant industrial food systems. Many early FSM statements contrasted food sovereignty with food security, and promoted the production of food crops and sales to local markets, often arguing against the international commodities trade.[3] Several recent publications question the dichotomies of food sovereignty (FS) vs security and food sovereignty vs international trade.[4] In summary, increased academic attention to FS has launched a generative dialogue with social movement leaders and highlighted the need for additional field research.

The conceptual and empirical relationships among fair trade, food sovereignty and food security are underexplored. In this paper I contribute to bridging this gap through a case study focused on the experiences of a smallholder fair trade coffee-exporting cooperative in northern Nicaragua's highlands. My aim is to explore what food sovereignty scholars and advocates could learn from an analysis of the governance debates in the fair trade system, the case study of an influential smallholder co-op, and potential linkages among these projects. The paper is organised around the following research questions: (1) Which institutional changes do a community of smallholders identify as generating the most significant influences on their autonomy and food security? (2) How have smallholders and cooperative staff members interpreted and practised fair trade, food sovereignty and food security? (3) What strategies did an influential fair trade cooperative employ to navigate the recent split between Fair Trade USA and Fairtrade International? (4) In this broader historical context, how significant are links to these contested global fair trade networks for efforts to achieve food security and food sovereignty?

The paper argues that the juxtaposition of fair trade, food security and food sovereignty as competing terms is more perplexing than helpful to policy dialogue on questions of farmer empowerment, hunger alleviation and agricultural sustainability in the global food system. Instead of analysing these terms as rival categories, I explore a relational interpretation and practice by analysing the case of a cooperative that must simultaneously navigate cleavages in the governance of the global fair trade system as well as attend to farmer demands for food security and sustainable livelihoods. In response to these questions, I develop the case study and then discuss lessons learned.

The question of trade in food sovereignty

In 2014 Burnett and Murphy published an important article analysing the ambiguous position that global trade holds in the food sovereignty agenda. After pointing out that 'trade remains important to the realization of the livelihoods of small-scale producers, including peasants active in the Food Sovereignty movement, [and that] it also matters for food security', they argue for the development of a more pro-international trade FSM strategy.[5] Although the FSM claims to work against a free trade system they characterise as dominated monopolistic corporate power, global commodity systems and agricultural dumping, less is said about the entangled relationships connecting global commodity systems and incipient food sovereignty alternatives in everyday practice.[6] The Nyéléni Declaration (2007) includes a short segment about trade, stating: 'Food sovereignty promotes transparent trade that guarantees just incomes to all peoples as

40

well as the rights of consumers to control their food and nutrition'.[7] Recent statements from FSM organisations have affirmed the importance of market access for smallholders and women, lobbied against multilateral trade agreements under negotiation (eg the Transatlantic Trade and Investment Partnership), and demanded changes in global trade policy to support food sovereignty, sustainable peasant agriculture and agrarian reform as the way to eradicate hunger.[8]

The FSM will continue engaging international trade debates but the strategic direction appears to be undecided. Burnett and Murphy's provocative suggestion that the World Trade Organization (WTO) and not the Food and Agriculture Organization (FAO) is the appropriate forum for this strategic engagement is beyond the scope of this paper. If the FSM chooses to engage global trade negotiations more directly, it will probably push national governments to use the WTO or FAO as a forum to support agreements similar to the recently announced US–India agreement at the WTO, which allows countries to maintain national food reserves for food security purposes.[9] This plan could influence the effectiveness of government-led attempts to address price volatility and the food security of millions (possibly tens of millions). More research is needed to understand how this type of national government-led food sovereignty strategy interacts with the community-based and civil society-led approaches to construct FS from the bottom up. This article complements Burnett and Murphy's global food policy analysis and advances the debate with nuanced producer and agrarian civil society perspectives. For example, their proposition that the FSM agenda is broadly consistent with fair trade does not clearly distinguish between the different versions of fair trade.[10]

Free trade vs fair trade

Like food sovereignty, fair trade originally emerged as a response to the exploits of the global free trade system. In contrast to FS, pioneering fair traders did not eschew global trade but sought to develop fairer North–South partnerships that enhanced market access and paid better prices to marginalised smallholder and artisans. Today scholars conceptualise fair trade in different ways, including as an ethical consumer marketplace linked to rural development projects, an ineffective neoliberal fantasy, and as an alternative food system working 'in and against' the dominant commodities market.[11] The globally accepted definition states that fair trade is a trading partnership based on dialogue, transparency and respect that seeks greater equity in international trade; it claims to prioritise smallholder and worker empowerment and requires that traders pay minimum prices and social development premiums to producers who meet established standards.[12] Historically fair trade was about both reforming free trade and eventually replacing it with an 'alternative' fairer system. The creation of a third-party certified market became the primary avenue for this work. Certified Fairtrade coffee is the mainstay of the international network of fair trade labelling organisations that have expanded rapidly since the development of a product-based certification approach in 1988. Fairtrade International now reports total retail sales for all products of US$7.3 billion, and 670,000 coffee smallholders affiliated with co-ops certified to export into these markets. Fair trade coffee accounts for about 3% of the global coffee industry.[13]

Questions about the governance of fair trade have evolved over the past 20 years, often in response to pragmatic debates about how to increase producer access through market growth, impact and inclusion. The early and persistent question asks: what are the impacts of fair trade among participating farmers?[14] Later, researchers questioned how fair trade could scale up without compromising its core values.[15] As larger corporations started to sell certified Fairtrade products, the markets expanded (now topping 1.4 million producers and workers) and academics increased their scrutiny of Fairtrade governance.[16] These governance questions became increasingly pressing when Fair Trade USA split from Fairtrade International.

'Big tent' Fairtrade

In the previous two decades smallholder farmers and food justice advocates collectivised power and used fair trade in an attempt to build a fairer and more sustainable market. Initially this market growth represented a tangible alternative to the low prices paid to producers in the anonymous commodity markets. Fair trade coffee, chocolate and other foods quickly became available as certifiers developed criteria that allowed mainstream firms, such as Starbucks and Nestlé, to sell 'Fairtrade'. In the early 2000s the low international prices in the coffee commodity markets made the price floor offered by certified Fairtrade markets a tangible, though limited, alternative model. The collaborative work of smallholders, advocates, cooperatives and businesses to launch and expand a global fair trade system sought to transform unfair trade relationships into 'a different kind of market', which empowers small-scale farmers, workers and consumers.[17]

From 2007 to 2011 tensions in the 'big tent' version of Fairtrade remained elevated as high commercial coffee prices coincided with declining economic returns for producers' sales through Fairtrade. At the same time Fairtrade's product quality requirements increased. A number of debates erupted, concerning the floor price, price premiums paid to farmer organisations, the possible inclusion of large coffee plantations, and the degree of power that organised farmers should have in governing the system (ie how many seats, if any, should representative smallholder organisations have on Fairtrade International's board of directors). There were also arguments about the aspirational goals of this system: is fair trade about increased market access to the existing global grade system, or is it about creating an alternative to a fundamentally unfair trade system?

The US-certified fair trade market has expanded vigorously since the Institute for Agriculture and Trade Policy launched TransFair USA as a certification and product licensing nonprofit organisation affiliated with the global network of fair trade labelling organisations then called FLO (later called Fairtrade International). TransFair USA hired Paul Rice as the founding CEO in 1998. After significant market growth in the early 2000s this organisation often critiqued its European counterparts over slower market growth and the high costs.

The split emerged when TransFair USA announced its separation from Fairtrade International, changed its name to Fair Trade USA, and launched 'Fair Trade for All' on 15 September 2011. Fair Trade USA promised to make fair trade flexible for business and to double sales by 2015.[18] Several key components of this proposal include, the development of a new standard for unorganised smallholders, the inclusion of large coffee plantations, and increasingly flexible standards that apply to Northern businesses.

Fair Trade USA vs Fairtrade International

Big tent fair trade collapsed as social movement organisations and many smallholder cooperatives became increasingly disillusioned with the mainstreaming strategy and undemocratic governance structure common to Northern certification agencies, particularly Fair Trade USA. The corporate model of Fairtrade offers less power to smallholders, since minimum prices have not kept up with the spiralling costs of sustainable production, and new large-scale entrants were able to usurp existing relationships with importers, roasters and retailers built by smallholder producer organisations and alternative trade allies. Hybrid approaches linking social economy-oriented cooperatives to more small- and medium-sized 'socially responsible' companies have provided smallholder organisations with expanding markets, increased access to credit and grassroots development projects, education and women's rights. The solidarity-oriented alternative trade model represents a third version of fair trade. It consists of 100% fair and alternative trade organisations in the North (eg worker cooperatives, such as Equal Exchange, and mission-driven, not-for-profit organisations, such as SEERV) and many, but not all, of the producer cooperatives in the South. These are the founding alternative trade organisations that launched their system more than 40 years ago.[19]

From a food sovereignty perspective an analysis of fair trade divisions suggests procedural justice questions about the lack of voice and vote for farmers, workers and consumers in governance and standards development. There are no seats designated for alternative trade groups on Fair Trade USA's board of directors. While producers have three seats on Fairtrade International's board, small-scale producers lack proportional representation. In the past five years Fairtrade International has made several significant governance changes, including making smallholders 50% owners of their global regulatory system and electing a smallholder representative from the Dominican Republic as the new president of their board of directors.[20] Despite recent reforms in this direction, Fair Trade USA's board of directors still does not include representation from smallholder fair trade cooperative networks or civil society.

Food security and the food sovereignty social movement

FSM organisations have historically defined FS in opposition to free trade strategies to achieve food security, and to the use of the term 'food security' itself.[21] There are multiple interpretations of food security and food sovereignty, and a dichotomy between the terms is misleading. Food security is a concept that describes the condition of access to adequate food, while the latter term is more explicitly a political agenda for how to address inadequate access to food and land rights in a way that simultaneously promotes sustainable peasant agriculture and farmer empowerment.[22] A relational approach between these terms opens the possibility of incorporating food security insights into FSM strategies. Key ideas underpinning the food access approach to food security are credited to Sen's entitlement theory, which explains how households use multiple relationships, including wage labour, other income sources, food production and exchange, to bring food into their house.[23] Like fair trade and food security, food sovereignty is also an evolving concept, a discourse, and an emerging set of alternative farming, fishing and food distribution practices. In a recent meeting of the Committee on World Food Security in Rome delegates from leading FSM organisation La Vía Campesina (LVC) continued to insist on food sovereignty as the primary strategy, but instead of opposing food security they called on states to improve food security outcomes.[24]

The largely rural membership of many organisations that formed the early stages of the FSM, which now includes many urban agriculture initiatives too, was spurred by the fact that more than two-thirds of food-insecure people live in rural areas.[25] The most common form of rural food insecurity is seasonal hunger. These lean months often correlate with harvest size, agricultural calendars, climate, higher staple food prices and the availability of rural employment and income.[26] Several frameworks analyse the causal forces that link poverty and hunger.[27] One explanatory framework analyses the historic political economic processes that shape the 'social spaces of vulnerability' to hunger, and focuses on hazards exposure, sensitivity and efforts to access food when crops fail or prices spike.[28] The social space of vulnerability changes over time as the historical political-economic forces influence the local institutions that shape household access to land, water, markets and credit, compromising their ability to cope with hazards and put food on the table.

Fair trade aims to empower producers and promote sustainable livelihoods, and FS aims to secure rights and eliminate hunger, but evaluative research is relatively sparse. Little published research empirically measures the local food security impacts and farmer empowerment processes among households and organisations affiliated with FSM organisations, although a recent publication examines if FS could happen among a group market-oriented smallholders in Indonesia.[29] Furthermore, researchers have not examined the intermingling of fair trade networks and food sovereignty concepts within smallholder cooperatives and farming communities. A growing body of research assesses the household food security outcomes among farmers linked to fair trade, organic and conventional commercial coffee markets. The findings from Latin America consistently show mixed results. Households selling a portion of their coffee to Fairtrade reported several tangible benefits (eg higher prices or increased access to credit and technical assistance) in some cases, but persistent food insecurity in most cases.[30] Another study found that farmers affiliated with co-ops receiving significant development assistance from corporations purchasing their fair trade coffee reported shorter periods of seasonal hunger from 2007 to 2013, although the attribution of these changes remains unclear.[31] A large-scale survey recently completed in Ethiopia and Uganda found sparse evidence for positive fair trade impacts on poverty reduction, and poor working conditions on certified coffee, tea and flower farms and processing plants.[32] The following case study considers an influential cooperative with a long history of selling a high percentage of its coffee to Fairtrade markets. The co-op and affiliated farmers have started to engage FS concepts, but it is not affiliated with the larger FSM.

Approach, methods and study site

This study is guided by an embedded single case study method.[33] The units of analysis include the fair trade system as it intersects with PRODECOOP, a primary-level cooperative, and affiliated farmers. There are several generalisable aspects and other potentially unique characteristics in this case. This case is generalisable since millions of smallholders in Central America and worldwide must navigate seasonal hunger, low levels of formal education, poor drinking water and sanitation infrastructure, and vulnerability to climate change.[34] These communities faced food price spikes from 2008 to 2010, and changes in coffee markets.[35] This case is of unique interest because PRODECOOP's general manager has a leading role in coordinating the organised smallholder resistance to Fair Trade USA's decision to split from Fairtrade International and because Fair Trade USA's CEO, Paul Rice, launched his career with this same cooperative.[36]

My involvement with PRODECOOP started in 2000 through evaluation work of a project to improve coffee quality, during the 1999–2005 coffee crisis. The farm and community-level research started with a community-based participatory action research project that united a university researcher (myself) with not-for-profit organisation staff (both a local Nicaraguan NGO and an international NGO) and PRODECOOP. The partnerships represent a shared production of knowledge, local empowerment and design strategies to reduce seasonal hunger, increase access to healthy food and promote sustainable agriculture among 1000 affiliated families.[37]

My qualitative data collection methods included focus groups, participatory workshops and participant observation during farmer field days and in meetings. I reviewed internal documents and reports from PRODECOOP and communicated with staff members. These approaches integrate international community development and evaluation work with reflexive ethnography and mixed methods.[38] I also combined interviews and several community-based mapping activities that are broadly similar to those of other research projects mapping common property with resource users.[39] Most key informant interviews, workshops and focus groups were conducted in July and December 2013. I collected data on the overall context through participant observation during July and December field visits from July 2009 to December 2013. Focus group and interview participants were recruited purposefully to include voices of men and women, old-timers, community leaders and young adults.

The study area includes regions identified by PRODECOOP staff as the most food insecure, located in Estelí, Madriz and Nueva Segovia. Although most of Nicaragua's coffee is produced in the Districts of Matagalpa and Estelí, the Segovias region included in this study has an outstanding reputation for coffee quality and significant threats associated with the anticipated impacts of climatic change. The nested case study focuses on a primary-level cooperative in the highlands of the Condega, Estelí. The terrain consists primarily of small mountains, mesas and hills, ranging from 700 to 1550 meters above sea level. There is a rainy season lasting from May to October, followed by a dry season. The vegetation consists primarily of tropical dry forests at lower altitudes and semi-humid and mixed oak and pine forests at higher altitudes. Most farmers in the study area produce a combination of cash crops (coffee), subsistence crops (corn and beans), fruit trees and occasionally tubers and vegetables.

PRODECOOP and fair trade

La Promotora de Desarollo Cooperativo de las Segovias (the Promoter of Cooperative Development in the Segovias, PRODECOOP) is the flagship Nicaraguan cooperative involved in fair trade. Most of the farmers and rural organisers who founded PRODECOOP trace part of their inspiration to the ideals of Nicaragua's 1979 popular revolution. After the Sandinista government lost the 1990 elections, a small group of young agricultural and social development professionals organised with small-scale coffee farmers struggling to survive the dramatic changes affecting rural Nicaragua. They initially formed an NGO called Colibri and then founded PRODECOOP in 1994. PRODECOOP sold its first coffee containers to alternative trade organisations in Europe and the USA (Equal Exchange). These alternative trade organisations would later link with others to create the international certified Fairtrade coffee system. Farmers risked part of their crop (and with it their vital income and food security) when they committed this coffee to PRODECOOP before receiving payment. The alternative trade organisations risked losses as they purchased coffee from a new organisation with no prior experience of exporting. This trust facilitated the birth of an alternative agri-food system that sought to distribute value more fairly, to increase transparency and to empower communities.

PRODECOOP is a dynamic, effective organisation and I attribute this to its farmer leadership, politically astute, capable and solidarity-oriented administrators, the high quality coffee produced and processed in this region, and support from allies involved in fair trade networks. PRODECOOP's commercialisation staff uses its reputation for quality, reliability and consistency to negotiate better prices from importers and roasters. The co-op's general manager now claims that this is more important than any certification. However, Fairtrade and organic certifications continue to matter and historically these standards informed the norms and practices within PRODECOOP, while also providing access to markets, price premiums and credit, and contributing to its efforts to secure millions of US dollars in funding for community development projects. PRODECOOP generally pays affiliated farmers coffee prices that are 20%–30% above national averages.

These complexities make it difficult to attribute specific benefits to Fairtrade markets and the associated international development networks. However, the vigorous debates that PRODECOOP's leadership made following changes to international Fairtrade standards demonstrate that fair trade was worth a fight. Furthermore, the co-op's annual reports also show that Fairtrade's social development premium (now at 0.20/lb of exported green coffee) totalled about $800,000 in 2012–13.[40] Grants and donations from coffee companies and allied non-profit agencies have been significant, often totalling over $1 million in a single year. These grants and donations fund women's rights, organic production, coffee quality improvements, food security and farm diversification projects. The Fairtrade price floor can make a significant difference when international commodity prices fall. In the early 2000s PRODECOOP paid farmers $100/sack of Fairtrade organic coffee vs the $40/sack paid by intermediaries selling to commercial markets. The significance of the aforementioned action strategies is best interpreted in a broader historical context offered by the experiences of the rural community.

Histories of conflict, hunger and uneven access

In response to the first research question, I conducted a detailed case study in a rural community where the majority of the households are currently affiliated with PRODECOOP. In focus groups and interviews, residents recounted their history from the 1960s to the present. The story follows changing access to food and land as residents transitioned from dependent farmworkers and sharecroppers labouring for a cattle-and-coffee plantation owner to armed combatants fighting to topple the Somoza dictatorship and the local plantation's *patrón*, who dominated their lives and usurped their rights. After the revolution in 1979, the national government seized the plantation through the agrarian reform programme. Initially the state formed and directed a co-op of local residents, but later the title was turned over to a locally managed co-op. As male residents became increasingly involved with the 1980s war against the 'counter revolution', gender roles started to change in some places. In the 1990s the wars ended, but food insecurity persisted. Pests and droughts, which often coincided with El Niño-Southern Oscillation (ENSO) years, led to crop failures and contributed to an annual average of three to five months of seasonal hunger. Local intermediaries controlled coffee and corn-market access. Thus, increased farmer sovereignty was not initially matched by substantial gains in food security. Seasonal hunger decreased from the 1970s to the 1990s but it persisted even after the local co-op started selling to fair trade markets affiliated with PRODECOOP in 1994 started exporting to Fairtrade coffee markets.

Residents agreed that the five most important common-pool environmental resources were: water sources (streams, creeks and springs); forests for water conservation, tree fruits and firewood; animal habitat; pastures for livestock; and seeds and genetic resources. The title for the *milpa* agricultural plots (for planting corn, beans, squash and other crops) was collectively held, but farmers managed this land individually. The next sections analyse changing patterns of access to these environmental commons, household food security, and farmer autonomy during the previous six decades.

The hacienda and dependency in the 1960s and before

Although elderly residents reported less violence than found in the armed appropriation of coffee land from indigenous communities in Matagalpa, they mentioned uneven and manipulated land deals, explaining how hacienda owners became *tierra tenientes* (landlords with large holdings) by offering local residents new clothes and cattle in exchange for their land titles.[41] One said, 'the indigenous people had a community focus but then the hacienda came and with it exploitation'. They also reported friendships between the hacienda owner and the Somoza dictators. The hacienda owners controlled access to forests and drinking water. There was rarely enough food in most households. Their uneven exchange entitlements tell a powerful story: 'We earned 4 *Córdobas* a day, and we needed to buy the corn from the company store on the hacienda at 10 *Córdobas* for 25 lbs'.

Violence, famine and seeds of resistance in the 1970s

With very limited access to natural resources and 80%–90% rural illiteracy rates, malnutrition was common, even though farmers reported consistent rains and were able to grow some food in sharecropping arrangements. Early childhood death was common, as one community experienced in the early 1970s: 'Families in the 16 houses of this community lost 30 children, nearly two per house from malnutrition compounded with diseases and illness, such as measles, diarrhoea and chickenpox'. The suffering associated with the cyclical periods of seasonal hunger deepened when drought slammed the community and led to local famine in the early 1970s, a crisis also reported in other northern coffee growing communities.

The hacienda dominated the local political economy during this decade, 'denying us our rights' and controlling access to forest firewood and fruits. Several community residents petitioned the hacienda owner, asking him to work with the government and open a school for the 30-plus school-age children. Although the law supported this right, the residents were promptly threatened and blacklisted, prompting them to join the armed resistance.

Agrarian reform, war and the state run cooperatives in the 1980s

Broad-based popular and armed resistance converged to topple the Somoza dictatorship on 19 July 1979, and the Sandinista government subsequently consolidated its power. The government seized resources controlled by Somoza and his cronies that accounted for up to 25% of the country's productive assets.[42] The hacienda passed into the hands of the new agricultural ministry. Initially the government managed land and labour. If a worker picked bananas and mangoes from the shade trees above the coffee, the manager (who in some ways was like a more humane *patrón*) would say nothing, 'but deductions for that picked fruit appeared on paychecks'. Food security and seasonal hunger remained a challenge but government donations were common, though not necessarily consistent. By the mid-1980s the government had transferred the title for 1540 hectares as a single commonly-held property from the state-run enterprise to this local cooperative. Once the new resident-led board of directors took control, residents could access the forest, including firewood, fruit and water, without permission or financial penalty. Some reported higher deforestation rates during this period; these reports are still unverified.

A limited peace, land loss and affiliation with PRODECOOP in the 1990s

Food access changed again with the end of the 1980s wars following the Sandinistas' electoral defeat in 1990. However, in October 1991 a group of re-armed rural residents attacked the cooperative, killing one local member and burning the existing infrastructure, including crop storage facilities, coffee and farm equipment. The goals of this attack are still unclear, but could have been linked to a large plantation owner's efforts to recapture land 'lost' during the agrarian reform.

During the early 1990s the cooperative's board initially organised agricultural labour collectively. Periods of seasonal hunger sometimes expanded beyond the expected June, July and August, and overall residents perceived that they were shorter than before. Residents also reported harvesting wild foods, such as *flor de izote* (*Yucca guatemalensis*), *hojas de bledo* (an uncultivated Amaranth) and fruits from the forests and coffee shade trees on the co-op's common-property forest. Many also obtained food in exchange for their weapons through government-backed pacification programmes in the path from swords to ploughshares.

Property divisions, empowerment and the persistence of seasonal hunger in the 2000s

Co-op-led collective action declined after the loss of 320 hectares of common property to a corrupt microfinance organisation that had invested in a cattle project with the cooperative's board of directors. In the mid-2000s the assembly of 93 members elected a new board of directors, and PRODECOOP later channelled international development funding from the coffee industry to this community for water projects, seed banks and farm diversification. Monthly cooperative meetings with no external agents present often drew over 50 participants, of whom at least 30% are female. In contrast to these collective efforts, the cooperative also continued to divide into individual private property rights. This occurred first in the area of basic grain production, but extended in the mid-2000s to coffee growing plots.

Calculations using geospatial data found that the cooperative lost 1.6% of its forest cover from 2000 to 2012 – better than the net change of a 5.9% loss in Nicaragua (based on the 2013 Global Forest Change project from the University of Maryland). Most of the deforestation in the cooperative occurred in order to plant corn and beans, to harvest firewood, and to sell trees for construction. More research is needed, but the transition from collective to individual land tenure may have contributed to deforestation in some places, and it probably limited the access of the poorest households to fruits and famine foods on previously commonly held property.

One focus group within the larger workshop asked four female participants to respond to the five thematic prompts and to record their experiences during the 1990s. They promptly deviated from the instructions and, after a thoughtful discussion, wrote their own consensus-based history. Their representative stood up and read the following statement, which I excerpt here: "We had no land. We worked on rented land. The poor life made it very difficult to study. As parents we were subordinated to the rich…Only those with money could pay a real teacher.' After getting the land title and joining the co-op, the narrative continues:

> First we have the land, fertilising the land, receiving training and [exercising our] power to make use of it, and to be able to respond to the lean months. Today we are free to study. Today we own our own plots…Today, as a woman, as a mother, I am free to speak, to decide and to make our own decisions, and we are no longer oppressed.

The broader history from the 1960s to the present illustrates the fundamental importance of agrarian reform in improving food and land access. This is a core part of an FSM strategy. The narrative also suggests the key role of a popular revolution, and the initial state-based intervention to create the co-op, which was a requirement for accessing usufruct rights on this collective land title. Finally, it shows how affiliation to PRODECOOP generated benefits such as support for land tenure, technical assistance for organic coffee production, and better coffee prices from sales to Fairtrade and organic markets. PRODECOOP also used fair trade-friendly coffee roasters and development assistance organisations to channel aid to this local cooperative, which supported building the co-op's offices, training for sustainable agriculture and a community-based grain bank.

Despite a willingness to invest in collective action supporting access to global coffee markets, local foods, drinking water, education, seeds and farm inputs, the local institutional history shows that most rural residents were dissatisfied with the forced collectivisation of agricultural land management during the early phases of the agrarian reform. Interviews with farmers and co-op staffers show that in this community the basket of individually held property rights has consistently expanded during the past 25 years. The broader political economic and local socio-cultural reasons for this pattern, which are common in peasant communities worldwide, are beyond the scope of this paper.

Local interpretations of food security and food sovereignty

Farmers and PRODECOOP staff members often interpreted food security and food sovereignty in similar ways and as complementary projects. However, the farmers emphasised empowerment and co-op staffers focused on FS methods, and not the FSM. One staffer stated: 'Food security is when there is enough food for a full meal for everybody, every day of the year. Food sovereignty means that we produce sufficient amounts of our own healthy food, preserve quality, and do not rely on external inputs.' Farmer perceptions tended to focus more on production and autonomy. Some farmers were unfamiliar with the term, but most conveyed that it was about the need to secure land tenure, freedom to choose which crops to plant, and their ability to save and share locally adapted seeds.

As an organisation PRODECOOP began integrating concepts from FS, fair trade and food security. In partnership with local and international NGOs it developed cooperative-led grain banks, which purchase, store and redistribute corn and beans to affiliated farmers and local residents. This strategy uses the cooperative's structure to manage these enterprises and apply fair trade principles to set prices, while prioritising food access. Farmers also participated in community-based seed banks. The idea for the seed banks came from participation in a farmer-to-farmer exchange, as many reported that seed banks increased their sense of sovereignty. However, more research is needed to analyse how participation in these seed-saving initiatives affects crops yields and seasonal hunger.

PRODECOOP's general assembly recently approved a comprehensive 10-year strategic initiative to promote food sovereignty and food and nutritional security. This plan references Nicaragua's 2009 food sovereignty law as its conceptual framework. When asked to explain the role of cooperatives in food security, one staffer said, 'Food security requires solidarity. We need it for soil conservation works, seed exchanges, and to make organic inputs. Many families are unable to do all of this on their own, but several families united together as a brigade can handle it. Furthermore, this strengthens the social ties as the community collectively learns the problem areas.' Although PRODECOOP's staff members promote many FS-related principles and express interest in the technologies (eg bio-fertilisers and local seed banks) used by FSM organisations, they stated that their alliances are with the 'cooperative movement'.

Smallholder cooperatives navigating Fairtrade International and Fair Trade USA

The Latin American and Caribbean Network of Smallholder Fair Trade Producers (known as the CLAC for its Spanish initials) is the largest voice for organised smallholders involved in certified Fairtrade, representing more than one million individuals.[43] PRODECOOP's executive director, Merling Preza, remains active in the CLAC leadership discussion as well as in those with the Fairtrade Foundation and Fairtrade International's newly formed US affiliate office, called Fairtrade America. In this capacity she frequently led the opposition to Fair Trade USA's split from Fairtrade International and the former's modification of standards to include large coffee plantations. Soon after Fair Trade USA's Paul Rice announced the split from Fairtrade International in September 2011 and launched 'Fair Trade 4 All', Preza spoke on behalf of the CLAC, saying, 'It hit us like a cold bucket of water'.[44] Subsequent statements show a fundamental difference in the interpretations of fair trade, specifically concerning issues of collective empowerment, Fair Trade USA's governance framework and its proposals to certify large coffee estates and unorganised smallholders (selling through private or transnational exporters). The CLAC also raised concerns about the lack of representation from organised smallholders on Fair Trade USA's board, and lauded Fairtrade International's decision to expand organised smallholder representation in its governance and ownership structures. Preza stated that PRODECOOP had refused to coordinate activities with Fair Trade USA for a while (18 December 2013) and then noted broader challenges:

> There has been an evolution as [fair trade] has practically passed from being a movement to being a certification that generates any quantity of rules. This is very distinct from when we were a movement for transparency and more direct relationships focusing on knowing the small-scale farmer and understanding why they live in their current situation in which the market does not compensate for their work. Fair trade was born of the necessity of the farmer who had only a small quantity of coffee and no access to the market. Things continued to change with increased access to the market, but all the changes in recent years have focused on verification, traceability, and the environment, this gave fair trade more focus on economic and environmental issues than social concerns (Merling Preza).

As political conflicts among fair trade certification agencies endured, most cooperatives adopted a pragmatic strategy to maintain existing market share and expanded their political advocacy. PRODECOOP continued exports to coffee roasters that supported Fair Trade USA's decision to split away from Fair trade International, supported reforms within Fairtrade International and initially disassociated itself from Fair Trade USA. For example, representatives from the CLAC celebrated when Nestlé signed an agreement Fairtrade International instead of Fair Trade USA. However, Preza remains concerned about unfair competition when Fairtrade-certified transnational firms, particularly exporters, undercut cooperatives, noting that '*si pones el ratón con el gato, el gato va comer el ratón*' (If you put the mouse and the cat together, the cat is going to eat the mouse). In the past three years Preza also shifted away from global market access and fair trade governance debates and started to focus PRODE-COOP's strategy on farmer livelihoods and food security.

Discussion and conclusions: lessons for food sovereignty and fair trade

This discussion synthesises six lessons learned. Cooperative staff and farmers have started to integrate ideas from fair trade, food sovereignty and food security to achieve their organisational and livelihood goals. The historical analysis of agrarian change and rural hunger shows the significant influence of the political-economic context, local institutions and agrarian reform as rural residents have increased their autonomy and access to food and land, advancing towards FS goals; however, improved food security has not immediate followed. The FSM should not overlook tools that enable fairer avenues of market access. However, the increasingly muted producer voice in certified fair trade governance debates suggests that mainstream versions of fair trade (eg Fair Trade USA) conflict with the FSM's normative goals. The creation and sustenance of capable and democratic smallholder cooperatives can advance FSM goals and improve food security outcomes. Under the right circumstances, access to Fairtrade markets could strengthen producer associations. However, the PRODE-COOP case could be unique given the history of the Nicaraguan revolution and significant support from international development organisations. Fair traders could learn from the oppositional politics, mobilisation strategies and tactics that the FSM regularly musters in defence of farmer autonomy. Finally, both fair trade and FS share the challenge of improving food security outcomes in the context of climate variability. Efforts to achieve this common goal could benefit from tactical and possibly strategic alliances for action.

Taken together, continuities in farmer struggles to increase their autonomy and access to food and natural resources, local interpretations of food security and sovereignty, and PRODECOOP's strategic action plan to address rural hunger demonstrate how co-ops and producers find synergies among these terms. The second point illustrated by this case is that increased autonomy is not necessarily followed by improved food security outcomes. After the agrarian reforms and affiliation with PRODECOOP, these community members report that livelihoods and gender equity improved. PRODECOOP supported the transition to certified organic coffee production, often selling over 80% of the local cooperative's coffee to Fairtrade markets, and it channelled international development assistance. Seasonal hunger persisted. Although considered less significant than the agrarian reform, residents concurred that these recent institutional changes and the increased market access favoured their livelihood goals.

The findings of this study concur with recent suggestions that the FSM consider advocating trade policy changes and complementing its critical analysis of smallholder commodity production with a benefits assessment.[45] However, my study of fair trade governance and local resistance indicates the need for greater attention to power imbalances and questions of representation when analysing the FSM's strategic options to influence agricultural trade negotiations, even in the certified market place. For example, the increasingly muted producer voice in Fairtrade governance exposes incompatibilities between Fair Trade USA and other mainstream approaches to this project and the FSM's political goals. However, alternative trade organisations, producer co-ops and regional smallholder networks, such as the CLAC, will find synergies with the FSM priorities focused on smallholder empowerment, rights, food access and agricultural biodiversity conservation.

The fourth lesson concerns the key role of effective local institutions in advancing the overlapping goals of all three projects, provided that these institutions support farmer empowerment, sustainable agriculture and food security. Although there are no panaceas, effective local intuitions (eg smallholder cooperatives) are often a key factor in the sustainable governance and management of commonly held environmental resources, such as water, biodiversity, forests and pastures.[46] Agricultural production and marketing cooperatives could also address the exchange, donation and production dimensions of food access.[47] Furthermore, co-ops are the basic unit for representing producer interests in agricultural, environmental and food policy debates. This highlights the need for investments to create dynamic, capable and accountable smallholder cooperatives.[48]

This case study shows how state-led agrarian reforms helped to create the primary cooperative in the 1980s and how this later intersected with the international fair trade networks and regional organising efforts that created PRODECOOP. Farmers reported several changes, including an increased sense of empowerment, a better coffee price and gradual food security improvements, although challenges persist. Deforestation rates on the cooperative's property are below national averages.

These experiences and the cases of other smallholder co-ops in northern Nicaragua have implications for fair trade standards and the FSM's agenda of supporting local institutions. Although the FSM frequently critiques the lack of democracy in global food policy making, it is not obvious how it proposes to strengthen democracy and operational capacity within its affiliated organisations. Producer unions, co-ops and associations have a more democratic structure than the corporate ownership patterns that LVC critiques, but many local institutions are fragile and local elites can usurp power. Farmers and scholars have also criticised the lack of fair representation in Fairtrade governance debates.[49] In contrast to the FSM, Fairtrade standards include criteria that explicitly claim to enhance smallholder co-op capacity (eg price premiums for social development that are allocated by a cooperative's general assembly). More research is needed to assess the degree to which links to Fairtrade markets and the associated inspection and certification processes influence the capacity of export cooperatives and other local institutions. The evidence thus far is mixed.[50] Multiple factors explain the rise of PRODECOOP and its persistent challenges, but history shows that Fairtrade has made a difference for both the organisation and the cooperative and farmers involved in this case study. However, this case could be exceptional in this respect and caution should be exercised when generalising to all 2300 households affiliated with PRODECOOP, elsewhere in Nicaragua, and certainly to other continents. The varied histories, configurations and politics of agricultural cooperatives in different contexts suggest that studies of agrarian

change, coffee quality and market access precede the attribution of local outcomes to certified markets.

Changes to fair trade standards raise questions about future impacts and governance. Fair Trade USA's new standard allows plantations to export certified coffee.
Fairtrade International does not permit this, although both allow transnational
companies to export into certified markets. A degree of competition for farmer participation among co-op exporters can stimulate bottom-up accountability through
competition for farmer coffee and membership. However, the entrance of large
plantations into Fairtrade value chains will confuse consumers and could stimulate
a race to the bottom as co-ops start to skip the costly meetings associated with
attempting to run a democratic association.[51] Depending on the local response, this
would probably decrease the effectiveness of local institutions and limit their
efforts related to food security or environmental management. Astute smallholder
cooperative managers and smallholder representatives, such as Preza, express frus-
tration with the lack of democratic governance and empowered partnership of the
mainstream fair trade model: 'I don't discount what Fair Trade USA and Paul have
done to build the market, but it is the way they think for us…I don't like it when
people think for me. I want us to think together.'

The historic experiences of resistance and agrarian reform lived by the smallholders in this study, and Preza's role representing both PRODECOOP and the
CLAC in the fight for a fairer fair trade, show that Burnett and Murphy under-
estimated peasant aspirations, when they stated that several examples 'demonstrate smallholder farmers resisting radical and ideological change and instead
looking for practical opportunities'.[52] Indeed, the smallholders and co-op leaders
interviewed in this study are interested in practical strategies (eg continuing to
sell their coffee at better prices to large companies, using new bio-pesticides, or
exploring alternative marketing channels), but most local leaders have developed
a sophisticated political analysis and they are arguably more likely to influence
an 'outsider''s ideology than to be unwilling subjects of external ideological
manipulation. Furthermore, many of these farmers recognise that radical changes
(eg land reforms, changing the structure of local agricultural and credit markets,
domestic violence prevention campaigns and female asset ownership) were
needed to achieve their current sense of empowerment, and more than a few are
still willing to struggle for a more transformative and revolutionary food system.

Can fair traders engage with the FSM to avoid a bait-and-switch as farmers,
advocates and consumers lose the power to influence Fair Trade standards and
meanings? The split is growing within fair trade, as movement-based organisations and smallholder organisations have become increasingly disillusioned with
the undemocratic governance structure among many Northern certification agencies, particularly Fair Trade USA. The diverse approaches to fair trade can be
mapped to different types of fair trade. While minimum prices fail to keep up
with the spiralling costs of sustainable production, and new large-scale entrants
usurp export platforms that smallholder producer organisations had painstakingly
developed, the corporate model of fair trade risks offering fewer benefits (eg
declining real minimum prices) and less power to smallholders. Caught in the
middle – and pulled in both directions – hybrid approaches have provided
smallholder organisations with increased access to credit and have supported
grassroots development projects.

After the split between Fairtrade International and Fair Trade USA representatives from PRODECOOP engaged in open conflict with their previous allies from Fair Trade USA. However, the negotiations with the coffee roasters who also backed these changes were partially constrained by the commercial sales and donations that the roasters provided to the co-op. This could be one reason that the co-ops never coordinated the types of oppositional protest commonly used by the FSM. Nor were rights-based demands made to government agencies. Although they launched their own competing certification system, it would potentially have been more effective to mobilise mass protests and file complaints with governments. These tensions among the radical, reformist and neoliberal approaches to food system change will continue.[53] Tactical alliances and method-sharing for practical action illustrate one way that the more reformist approach of many fair trade enterprises could engage the FSM's more radical agenda.

Fairtrade, FS and other strategies for global food system change face the challenge of improving food security outcomes, reducing farmer and farm worker marginalisation, and conserving the environment. The persistence of rural hunger and the failure of mainstream trade and development projects contributed to the rise of the FSM's increasingly high-profile efforts to transform food systems. Although an initial read might identify synergies between fair trade and food sovereignty approaches to changing global food systems, the split within fair trade shows that a careful assessment of the ethics, organisational models and the many governance structures should precede the formation of strategic alliances. Both approaches aim to strengthen dynamic smallholder cooperatives, but not all stakeholders or versions of fair trade share the FSM's commitment to smallholder empowerment and sustainable agriculture. All strategies will also be tested based on their ability to eliminate hunger, secure human rights and sustain diverse agricultural ecologies.

Acknowledgements

Thanks to PRODECOOP's farmers and staff, and Community Agroecology Network consultant Maria Eugenia Flores Gomez. David Beezer, Alexandra Cabral, William Burke and Rica Santos provide research assistance. I am grateful to the special issue editors and two anonymous peer reviewers for useful comments. Jun Borras and Annie Shattuck encouraged me to finish the Yale conference paper, as I navigated my grandfather Allen Bacon's passing. This paper is dedicated to Allen and his legacy as one who campaigned for peace.

Notes

1. "Food Sovereignty: A Critical Dialogue." Conference held at Yale University, September 14–15, 2013. http://www.yale.edu/agrarianstudies/foodsovereignty/index.html.
2. La Vía Campesina (LVC), "Main Issues 2014."
3. Patel, "Food Sovereignty."
4. Agarwal, "Food Sovereignty, Security and Democratic Choice"; and Clapp "Food Security and Food Sovereignty."

5. Burnett and Murphy, "What Place for International Trade in Food Sovereignty?," 1065.
6. LVC, "Stop the Free Trade Agreements."
7. Declaration of Nyéléni.
8. LVC, "CFS in Rome."
9. Murphy, "What you Need to Know about the India–US Agreement," November 20, 2014, http://www.iatp.org/blog/201411/what-you-need-to-know-about-the-india-us-agreement-at-the-wto#sthash.N6Z39kfS.dpuf.
10. Burnett and Murphy, "What Place for International Trade?"
11. Walton, "What is Fair Trade?"; and Fridell, "Fair Trade Slippages."
12. See Fairtrade International's website, http://www.fairtrade.net/.
13. Ibid.
14. Ibid.
15. Raynolds, *Poverty Alleviation through Participation*; and Jaffee, *Brewing Justice*.
16. Bacon, "Who decides what is Fair?"
17. Ibid; Raynolds, "Mainstreaming Fair Trade"; Jaffee, *Brewing Justice*; and Wilson and Curnow, "Solidarity ™."
18. Fair Trade USA, "Fair Trade USA Resigns."
19. VanderHoff Boersma, "The Urgency and Necessity of a Different Type of Market."
20. Jaffee, *Brewing Justice*.
21. Fairtrade International, "Fairtrade Producer elected Board Chair."
22. Jarosz, "Comparing Food Security and Food Sovereignty."
23. Clapp, "Food Security and Food Sovereignty." See also Patel, "Food Sovereignty."
24. Sen, *Poverty and Famines*.
25. FAO, "The State of Food Insecurity."
26. Devereux et al., *Seasons of Hunger*.
27. Watts and Bohle, "The Space of Vulnerability."
28. Bacon et al., "Explaining the 'Hungry Farmer Paradox'."
29. Li, "Can there be Food Sovereignty Here?"
30. Mendez et al., "Effects of Fair Trade."
31. Caswell et al., *Revisiting the 'Thin Months'*.
32. Cramer et al., *Fairtrade, Employment and Poverty Reduction*.
33. Yin, *Applications of Case Study Research*.
34. Jha et al., "Review of Ecosystem Services."
35. Bacon et al., "The Hungry Farmer Paradox."
36. Denaux and Valdivia, *Historia de PRODECOOP*.
37. Bacon et al., "The Hungry Farmer Paradox."
38. Mikkelsen, *Methods for Development Work*.
39. Bauer, "On the Politics and Possibilities of Participatory Mapping."
40. PRODECOOP annual reports.
41. Austin et al., "Role of the Revolutionary State."
42. Agarwal, "Food Sovereignty, Food Security and Democratic Choice."
43. Coscione, "La CLAC y la Defensa del pequeno productor"; and CLAC/Fairtrade, "Comercio Justo."
44. "Merling Preza makes the Case against FT4all." Interview with Michael Sheridan, November 9, 2011. http://coffeelands.crs.org/2011/11/merling-preza-makes-the-case-against-ft4all.
45. Burnett and Murphy, "What Place for International Trade?"
46. Dietz et al., "The Struggle to Govern the Commons"; and Agarwal, "Food Sovereignty, Food Security and Democratic Choice."
47. Bacon et al., "Explaining the 'Hungry Farmer Paradox'"; Agarwal, "Food Sovereignty, Food Security and Democratic Choice"; and Sen, *Poverty and Famines*.
48. Poole and Donova, "Building Cooperative Capacity."
49. Jaffee, *Brewing Justice*.
50. Ibid.
51. Holt-Giménez and Shattuck, "Food Crises, Food Regimes, and Food Movements."
52. Burnett and Murphy, "What Place for International Trade in Food Sovereignty?," 1072.
53. Holt-Giménez and Shattuck, "Food Crises, Food Regimes, and Food Movements."

Bibliography

Agarwal, Bina. "Food Sovereignty, Food Security and Democratic Choice: Critical Contradictions, Difficult Conciliations." *Journal of Peasant Studies* 41, no. 6 (2014): 1247–1268.
Bacon, Christopher M. "Who Decides what is Fair in Fair Trade? Agri-environmental Governance of Standards, Access and Price." *Journal of Peasant Studies* 37, no. 1 (2010): 111–147.
Bacon, Christopher M., William A. Sundstrom, María Eugenia Flores Gómez, V. Ernesto Méndez, Rica Santos, Barbara Goldoftas, and Ian Dougherty. "Explaining the 'Hungry Farmer Paradox': Smallholders

and Fair Trade Cooperatives Navigate Seasonality and Change in Nicaragua's Corn and Coffee Markets." *Global Environmental Change* 25 (2014): 133–149.

Bauer, Kenneth. "On the Politics and the Possibilities of Participatory Mapping and GIS: Using Spatial Technologies to Study Common Property and Land Use Change among Pastoralists in Central Tibet." *Cultural Geographies* 16, no. 2 (2009): 229–252.

Burnett, Kim, and Sophia Murphy. "What Place for International Trade in Food Sovereignty?" *Journal of Peasant Studies* 31, no. 6 (2014): 1065–1084.

Caswell, M., V. E. Méndez, M. Baca, P. Läderach, T. Liebig, S. Castro-Tanzi, and M. Fernández. *Revisiting the 'Thin Months' – A Follow-up Study on the Livelihoods of Mesoamerican Coffee Farmers.* CIAT Policy Brief 19. Cali, Colombia: Centro Internacional de Agricultura Tropical (CIAT), 2014.

CLAC/Fairtrade. "Comercio Justo como una Herramienta para el Desarrollo Sostenible." Presentation, November 18, 2014, Managua.

Clapp, Jennifer. "Food Security and Food Sovereignty: Getting past the Binary." *Dialogues in Human Geography* 4, no. 2 (2014): 206–211.

Coscione, Marco. *La CLAC y la Defensa del pequeño Productor.* República Dominicana: Funglode, 2012.

Cramer, Christopher, Deborah Johnston, Carlos Oya, and John Sender. *Fairtrade, Employment and Poverty Reduction in Ethiopia and Uganda.* London: TEPR, 2014.

"Declaration of Nyéléni." 2007. http://www.nyeleni.org/spip.php?article290.

Denaux, Guillermo, and Salatiel Valdivia. *Historia de PRODECOOP.* Estelí, Nicaragua: PRODECOOP, 2012.

Devereux, S., B. Vaitla, and S. H. Swan. *Seasons of Hunger: Fighting Cycles of Quiet Starvation among the World's Rural Poor.* London: Pluto Press, 2008.

Dietz, Thomas, Elinor Ostrom, and Paul C. Stern. "The Struggle to Govern the Commons." *Science* 302, no. 5652 (2003): 1907–1912.

FAO, WFP, IFAD. *The State of Food Insecurity in the World 2012. Economic Growth is Necessary but not Sufficient to Accelerate Reduction of Hunger and Malnutrition.* Rome: FAO, 2012.

Fairtrade International. "Fairtrade Producer elected Board Chair February 2014." http://www.fairtrade.net/single-view±M582f64eee07.html.

Fair Trade USA. "Fair Trade USA resigns Fairtrade International–FLO membership." September 15, 2011. http://fairtradeusa.org/press-room/press_release/fair-trade-usa-resigns-fairtrade-international-flo-membership.

Fridell, Gavin. "Fair Trade Slippages and Vietnam Gaps: The Ideological Fantasies of Fair Trade Coffee." *Third World Quarterly* 35, no. 7 (2014): 1179–1194.

Holt Giménez, Eric, and Annie Shattuck. "Food Crises, Food Regimes and Food Movements: Rumblings of Reform or Tides of Transformation?" *Journal of Peasant Studies* 38, no. 1 (2011): 109–144.

Jaffee, Daniel. *Brewing Justice: Fair Trade Coffee, Sustainability and Survival.* Berkeley, CA: University of California Press, 2014.

Jarosz, Lucy. "Comparing Food Security and Food Sovereignty Discourses." *Dialogues in Human Geography* 4, no. 2 (2014): 168–181.

Jha, Shalene, C. M. Bacon, S. M. Philpott, R. A. Rice, V. E. Méndez, and P. Läderach. *A Review of Ecosystem Services, Farmer Livelihoods, and Value Chains in Shade Coffee Agroecosystems. In Integrating Agriculture, Conservation, and Ecotourism: Examples from the Field*, edited by B.W. Campbell and S. Lopez-Ortiz, 141–208. New York: Springer, 2011.

La Vía Campesina (LVC). "CFS in Rome: The Majority of Governments remain Blind to the Challenges of Food Security." October 17, 2014. http://viacampesina.org/en/index.php/main-issues-mainmenu-27/food-sovereignty-and-trade-mainmenu-38/1684-cfs-in-rome-the-majority-of-governments-remain-blind-to-the-challenges-of-global-food-security.

LVC. "Main Issues 2014." Accessed October 12, 2014. http://viacampesina.org/en/index.php/main-issues-mainmenu-27

LVC. "Stop Free Trade Agreements." Accessed October 12, 2014. http://www.viacampesina.org/en/index.php/actions-and-events-mainmenu-26/stop-free-trade-agreements-mainmenu-61/1683-stop-ttip-ceta-and-other-destructive-trade-policies

Li, Tania Murray. "Can there be Food Sovereignty Here?" *Journal of Peasant Studies* ahead-of-print (2014): 1–7. doi:10.1080/03066150.2014.938058.

Mendez, V. Ernesto, Christopher M. Bacon, Meryl Olson, Seth Petchers, Doribel Herrador, Cecilia Carranza, Laura Trujillo, Carlos Guadarrama-Zugasti, Antonio Cordon, and Angel Mendoza. "Effects of Fair Trade and Organic Certifications on Small-scale Coffee Farmer Households in Central America and Mexico." *Renewable Agriculture and Food Systems* 25, no. 3 (2010): 236–251.

Mikkelsen, B., *Methods for Development Work and Research: A New Guide for Practitioners.* Thousand Oaks, CA: Sage, 2005.

Patel, Raj. "Food Sovereignty." *Journal of Peasant Studies* 36, no. 3 (2009): 663–706.

Poole, Nigel, and Jason Donovan. "Building Cooperative Capacity: The Specialty Coffee Sector in Nicaragua." *Journal of Agribusiness in Developing and Emerging Economies* 4, no. 2 (2014): 133–156.

PRODECOOP. *Informe Annual del ciclo 2012–13.* Esteli: PRODECOOP, 2013.

Raynolds, Laura T. "Mainstreaming Fair Trade Coffee: From Partnership to Traceability." *World Development* 37, no. 6 (2009): 1083–1093.

Raynolds, Laura T. *Poverty Alleviation through Participation in Fair Trade Coffee Networks: Existing Research and Critical Issues*. New York: Ford Foundation, 2002.

Sen, Amaryta. *Poverty and Famines: An Essay on Entitlements and Deprivation*. Oxford: Clarendon Press, 1981.

VanderHoff Boersma, F. "The Urgency and Necessity of a Different Type of Market: The Perspective of Producers organized within the Fair Trade Market." *Business Ethics* 89, no. 1 (2009): 51–61.

Walton, Andrew. "What is Fair Trade?" *Third World Quarterly* 31, no. 3 (2010): 431–447.

Watts, Michael J., and Hans G. Bohle. "The Space of Vulnerability: The Causal Structure of Hunger and Famine." *Progress in Human Geography* 17, no. 1 (1993): 43–67.

Wilson, Bradley R., and Joe Curnow. "Solidarity™: Student Activism, Affective Labor, and the Fair Trade Campaign in the United States." *Antipode* 45, no. 3 (2013): 565–583.

Yin, R. K. *Applications of Case Study Research*. Thousand Oaks, CA: Sage, 2011.

Food sovereignty and the quinoa boom: challenges to sustainable re-peasantisation in the southern Altiplano of Bolivia

Tanya M. Kerssen

Institute for Food and Development Policy/Food First, Oakland, USA

In the last three decades, quinoa has gone from a globally obscure food to an internationally traded product with rising global consumer demand. This transformation has had complex social and ecological impacts on the indigenous agropastoral communities of the southern Altiplano region of Bolivia. This article analyses the role that global quinoa markets have played in the repopulation and revitalisation of this region, previously hollowed out by out-migration. Yet, it also points to a number of local tensions and contradictions generated or magnified by this process, as peasants struggle to harness the quinoa boom as a force of 'sustainable re-peasantisation' and 'living well'. Finally, the article suggests that the food sovereignty movement should place greater emphasis on examining the culturally and histori-cally specific challenges facing re-peasantisation in particular places.

The southern Altiplano of Bolivia, once dominated by transhumant pastoral pop-ulations, is now experiencing a dramatic expansion of its agricultural frontier. As a result, the region is seeing a number of social, economic and ecological transformations. This expansion is the result of peasant-led efforts in the 1980s to forge global alliances and build an export market for quinoa at a time when neoliberal policies, combined with postcolonial perceptions of indigenous foods, made accessing domestic markets all but impossible. Their success generated an important livelihood opportunity in a long marginalised region marked by pov-erty and out-migration. Nonetheless, the rapid expansion of quinoa and the entry of new actors have engendered extractivist tendencies that threaten both the eco-logical sustainability and social integrity of agropastoral systems. Quinoa pro-ducers – as well as their trading partners, NGO allies, policy-makers and consumers – now find themselves at a crossroads, debating the path to a socially and environmentally sustainable future for the quinoa sector.

As global demand grows, cultivation expands to new frontiers and pressures on productive resources increase, the traditional custodians of the 'golden grain of the Andes' face an uncertain future.[1] How are Bolivian producers confronting this uncertainty? Is the development of the quinoa sector likely to contribute to 're-peasantisation' and local well-being in a sustainable way? And what lessons might the food sovereignty movement draw from this case? In order to address these questions, this article employs historical and political economic analysis of quinoa in the southern Altiplano; participant observation in the region; attendance at two international quinoa research conferences in Bolivia and the USA; and 17 semi-structured interviews with diverse actors in the quinoa sector in Bolivia and the USA, conducted between March and July 2013.[2]

Food sovereignty and re-peasantisation

A concept first popularised by the international peasant confederation, La Vía Campesina, at the 1996 World Food Summit, 'food sovereignty' is defined as 'the right of nations and peoples to control their own food systems, including their own markets, production modes, food cultures, and environments'.[3] As Desmarais explains, food sovereignty is explicitly rooted in the assertion of a peasant identity in the face of neoliberal capitalism, which declares the disappearance of the peasantry to be an inevitability of progress.[4]

The (re)affirming of peasant cultures and economies – or re-peasantisation – thus appears as a strategic necessity for the building of food sovereignty, particularly since 54% of the global population now lives in cities.[5] Indeed, the call for food sovereignty emerges at a seemingly dismal historical moment for peasants. Araghi, for instance, described massive urbanisation from 1945 to 1990 as a process of 'global depeasantization', in which Third World peasantries lost access to their means of subsistence and became rapidly concentrated in urban areas.[6]

A number of more recent analyses, however, have drawn more complex conclusions about the fate of the peasantry. Kay, for instance, has suggested that today 'the situation is more fluid and varied: not only do peasants move to cities, but urban inhabitants move to rural areas', generating what he calls a 'new rurality'.[7] Going even further, some scholars suggest that neoliberal globalisation has actually led to a *strengthening* of peasant identity – particularly in Latin America – through the emergence of peasant and indigenous social movements. Radcliffe, for instance, shows how indigenous peasant confederations in Ecuador began reclaiming indigenous dress in the 1990s along with other cultural and political strategies that strengthened 'Andean, rural, and agricultural identities'.[8]

For others, however, such as Bernstein, it makes little sense to talk about modern 'peasants' as a social category since, he argues, most if not all peasants have essentially 'become petty commodity producers, who have to produce their subsistence through integration into wider social divisions of labour and markets'.[9] The response of agrarian scholars in the pro-peasant or 'populist' camp, such as Van der Ploeg, has been to affirm that peasants are incorrectly understood as purely subsistence-oriented and disconnected from the wider (capitalist) world. Rather, 'peasants, their livelihoods, and their processes of production are constituted through the structure and dynamics of the wider social formation in which they are embedded'.[10] For Van der Ploeg, one of the defining features of the modern peasantry is its 'fight for autonomy and survival in a context of deprivation and dependency', a struggle he characterises as 'repeasantization'.[11]

Re-peasantisation, for Van der Ploeg, must not only involve a return to the countryside by non-peasants or former peasants, but also a return to 'peasant values' among the world's farmers. As Van der Ploeg explains, this implies a 'double movement':

> It entails a *quantitative* increase in numbers. Through an inflow from outside and/ or through a reconversion of, for instance, entrepreneurial farmers into peasants, the ranks of the latter are enlarged. In addition, it entails a *qualitative* shift: autonomy is increased, while the logic that governs the organization and development of productive activities is further distanced from the markets.[12]

In the decidedly peasantist food sovereignty literature, this qualitative shift generally involves – as Van der Ploeg lays out here – a distancing from markets and, ostensibly, a return to more subsistence-oriented production.

And yet, as Burnett and Murphy show, numerous prominent farmers' organisations associated with La Vía Campesina and the food sovereignty movement are engaged in the production of commodities – for export as well as domestic markets. They include ROPPA in West Africa and the National Family Farm Coalition in the USA.[13] As these authors argue, while 'the food sovereignty movement is identified with a strong preference for local markets', this tendency risks overlooking how peasants have, in the face of adverse local market conditions, utilised export markets as a strategy to remain on the land (and thus avoid the fate of urban migration).

In a recent critique of food sovereignty, Bernstein argues that food sovereignty advocates frequently use 'emblematic instances' of peasant practices (eg diversified and agro-ecological hillside production in Central America) that highlight the 'virtues of peasant/small-scale/family farming as *capital's other*'.[14] Similarly the term 'community' in food sovereignty discourse often 'exemplifies a "strategic essentialism" (Mollinga 2010), as in populist discourse more widely, which obscures consideration of contradictions within "communities"'.[15] Although I am not as willing as Bernstein to discard the term 'peasant', this paper seeks to apply greater scrutiny to the 'peasantry' and 'peasant community' in a particular place, highlighting some of the local tensions and contradictions at play in a peasant population that is far from homogeneous in its farming practices and in its position *vis-à-vis* capital. In so doing, I also suggest that the food sovereignty movement should place greater emphasis on recognising – as opposed to obfuscating – these tensions in the interest of advancing its political project.

The paper uses the term 'peasant' not as a fixed analytical category *per se*, but rather as a deliberately messy term that embodies, following Van der Ploeg, a continuum or 'grey zone' where processes of re-peasantisation and de-peasantisation are contested. I also analyse the complex role that global markets have played in facilitating re-peasantisation in the southern Altiplano of Bolivia on the one hand, and threatening its long-term viability on the other. While seeking to understand how global markets have affected Bolivian peasants, I also aim to analyse how peasants have affected markets. As Van der Ploeg observes, 'just as capital impacts upon the peasantries, the peasantries impact upon capital'.[16]

Lastly, this case highlights the importance of historically grounded, place-based analyses of peasantries 'under construction' and the challenges they face. While our theorisations need not be 'prisoners of place', in the words of Bebbington and Batterbury, analyses that theorise outward from cases can 'enrich and nuance our understandings of the intersections between globalization and contemporary rural life'.[17] Thus I begin by reviewing the social and historical context of food systems in the southern Altiplano. Next I discuss the transformation of quinoa from a globally obscure food disdained in national markets to a globally traded product with rising global (and to some extent domestic) consumer demand. In the third section I discuss the challenges that may impede a sustainable re-peasantisation. Finally, I address some of the ways in which Bolivian peasants are struggling to harness the quinoa boom as a force of re-peasantisation and 'living well' in the region.

Ayllus, autonomy and de-peasantisation in the southern Altiplano

Drastic climatic variations over short distances characterise the Bolivian landscape: from the semi-arid to arid cordillera and Altiplano in the West to the humid eastern mountain slopes and tropical rainforests to the East. For millennia politically independent pastoral societies traversed the North–South corridor of the Altiplano with large pack trains of llamas exchanging both ideas and products – such as salt, meat and fibre for potatoes, vegetables, coca and fish – with farming and fishing villages. The relationships developed by pastoralists with their sedentary trading partners became a form of kinship known as the *ayllu* which persists to this day – though greatly transformed.

Over time this movement of people, goods and genetic material among different ecological zones generated an extraordinary number of domesticated food crops and animals produced in non-contiguous territories, exploiting numerous ecological niches,[18] a system famously described by anthropologist John V Murra as a 'vertical archipelago'.[19] The high Andean plain (*Altiplano*) developed vital subsistence crops including tubers such as potatoes, oca (*Oxalis tuberosa*), and isaño (*Tropaeolum tuberosum*), and protein-rich 'pseudo-cereals' such as quinoa (*Chenopodium quinoa*), kañawa (*Chenopodium pallidicaule*) and kiwicha (*Amaranthus caudatus*). Of these crops, quinoa was particularly well suited to areas with high climatic risk such as the southern Altiplano – being able to withstand conditions of drought, salinity, wind, hail and frost in which other crops would perish.[20]

Risk management and dietary diversity in Andean food systems went hand in hand with the *ayllu* system, which were based on reciprocity relations; seasonal migration to various productive zones; communal resource management; and long-distance trade to exchange products from different regions and elevations.[21] Under this system, 'indigenous pastoral production was able for centuries to maintain a balance between demographic constraints and resource scarcity'.[22]

The Spanish conquest of the 16th century, however, radically disrupted this system. Confused by the *ayllu*'s discontinuous landholdings, Spanish administrators resettled Andean inhabitants into centrally located villages within bounded, contiguous territories.[23] While Spanish haciendas took over the best land, semi-autonomous Andean *ayllus* – now mostly severed from their extra-territorial linkages – were allowed to persist on the most remote lands, seen as unfit for agricultural development. Thus, with low rainfall (110–250 mm annually), more than 200 frost days per year, and generally poor soils, the southern Altiplano remained largely beyond the reach of the Spanish hacienda system. As Healy notes, "the *ayllu*'s territorial control became limited to mostly remote herding communities whose pastoral economies had little appeal for the landed oligarchy".[24]

In 1952 a social revolution succeeded in abolishing the hacienda system and redistributing land to thousands of highland peasants. However, agrarian reform did little to transform the southern Altiplano, where there had been few haciendas. By the 1970s the military dictatorships turned their focus towards reconstituting the agrarian elite in the eastern lowlands. As the country increased its production of lowland commodities like sugarcane and soy, US food imports also increased, transforming patterns of domestic consumption and creating a preference for wheat products, such as the now ubiquitous '*fideos*' (pasta) and white bread.[25]

The liberalisation of the economy in the 1980s further marginalised peasant food production as the terms of trade for highland, peasant-produced crops like potatoes, onions and barley rapidly eroded. Regional trade agreements such as the Tariff Union of the Andean Community of Nations (CAN) and agreements with Chile and Mercosur left peasants without protection from imports, bringing down the price of their products.[26] Farm incomes lost an estimated 50% of their purchasing power between 1985 and 1998.[27]

As throughout the global South, neoliberal restructuring spurred a dramatic wave of rural out-migration in Bolivia, mainly to the cities and to foreign countries like Chile or as far as Europe. The severe El Niño-induced drought that hit Bolivia between 1982 and 1984 also contributed to depopulating the countryside, triggering a 'migration explosion' out of the southern Altiplano.[28] Then, just as people were returning to their communities after the drought, the government introduced structural adjustment policies that not only removed protections for peasant agriculture, but also dismantled the state-owned mining sector – two primary rural livelihood strategies – leading to further rural depression and depopulation of the region.[29]

Although postcolonial marginalisation led to the impoverishment and depopulation of the southern Altiplano, it also remained a remarkable space where autonomous cultural, political and productive forms were able to persist. Long-distance trade and articulation with distant markets are not a novel occurrence in this region. Indeed, they are part of a livelihood strategy that predated the Spanish conquest and a repertoire of indigenous peasant resistance and adaptation.

'Quinoa re-peasantisation' through alternative global food networks

Food sovereignty and re-peasantisation generally assume a dynamic of 'localisation', retreat from (global) markets, and 'local production for local consumption'.[30] Contrary to these assumptions, this section argues that peasant-led efforts in the southern Altiplano in the 1980s led to *greater* market integration and helped to unleash a process of re-peasantisation linked to alternative global food networks. As Goodman et al note, the 'new politics of food provisioning' opened up by fair trade in the 1980s and 1990s 'build on imaginaries and material practices infused with different values and rationalities that challenge instrumental capitalist logics and mainstream worldviews'.[31] Nevertheless, these 'alternative' global trade networks are not immune to destructive market forces – challenges that will be address in the following section.

Quinoa's original expansion was made possible in part by the introduction of tractors to the southern Altiplano in the 1960s and 1970s, which brought the subsistence crop down from hillside terraces to the flat scrublands, previously reserved for grazing. While the state focused primarily on industrialising agriculture in the tropical lowlands during this period, some agricultural modernisation credits were extended to highland peasants to purchase tractors and disk ploughs.[32] NGOs and religious groups also promoted mechanisation in the Altiplano. Belgian missionaries, for instance, established a tractor-rental service in the province of Nor Lípez.[33]

When the Belgians left in 1975, they turned over the assets and management responsibilities of the project to local communities organised as a new cooperative entity called the Central de Cooperativas Operación Tierra (CECAOT). The National Association of Quinoa Producers (ANAPQUI) was later created in 1983, and the two organisations became the country's leading producers' associations growing and marketing *quinua real* ('royal quinoa') – a large-grained quinoa ecotype grown along the shores of the Uyuni and Coipasa salt flats (in what's known as the inter-salt flat or *intersalar* region) – which has since become the most prized quinoa on the global market for its large, white seed (> 2.2 mm) and high nutritional value.[34]

During the economic crisis of the 1980s – and particularly after the privatisation of state mines in 1986, which led to the lay-off of thousands of workers – miners relocated to the cities or to the tropics to plant coca. Others returned to their native communities in the southern Altiplano to grow quinoa.[35] These dynamics coincided with the growth in demand from the global North for speciality fruits and vegetables, organic products and health foods, which unleashed the non-traditional agricultural export (NTAE) boom in the global South.[36] In this context the demand for Andean quinoa products has grown precipitously, reaching 36 quinoa-importing countries in 2012.

With little external support CECAOT formed its own committee for industrialising quinoa processing, sending representatives to Peru to seek out new technologies and eventually building its own quinoa de-husker based on a barley-hulling machine.[37] Similarly ANAPQUI members worked tirelessly to improve processing methods, even travelling to Brazil carrying sacks of quinoa with them to test out rice and soy processing machines.[38]

CECAOT started exporting *quinua real* on a small-scale to the US-based Quinoa Corporation in 1984 – a company that pioneered the opening of the quinoa market in the USA. One of the company's goals was to revalue quinoa as a neglected food crop, not only in the USA, but also in its place of origin:

For the founders of the Quinoa Corporation, this was a necessary step that would eventually contribute to the food security of poor Bolivians, subjected as they were to a nutritionally inferior dietary regime based on highly-subsidised wheat products through US food aid. They hoped to increase internal demand and sales of *quinua real*, while at the same time contribute to improving the incomes and quality of life for indigenous producers of the southern Altiplano.[39]

Currently, 23.7% of Bolivia's quinoa production is sold in the domestic market, compared to 51.9% exported through legal channels and almost a quarter (24.4%) leaving the country as contraband.[40] While domestic consumption is said to have tripled between 2009 and 2013 – from 0.35 to 1.11 kg per capita[41] – this a small portion of domestic cereals consumption, which remains heavily dominated by wheat. According to the Food and Agriculture Organization (FAO), in 2009 Bolivians consumed 125.14 kg of cereals per capita, of which 45% consisted of wheat.[42] Further, 68% of the domestic wheat supply was imported – the legacy of an acute structural dependence on US food aid.

The emergence of quinoa as a globally traded crop in the 1980s and 1990s was arguably paramount in ensuring peasants' reproduction on the land in the southern Altiplano. This occurred at the height of neoliberalism, which was eroding livelihood options, especially in the countryside. There is also evidence that quinoa producers have fared better under subsequent crises because of their link to alternative global food markets. Pérez et al, for instance, indicate that quinoa farmers were better able to manage rising prices during the 2008 food crisis than producers of other crops such as potatoes.[43]

The point here is not to argue that neoliberalism *benefited* the southern Altiplano, but rather, to recognise the tremendous – unlikely even – achievement that is the contemporary quinoa sector. In the context of hostile neoliberalism peasants of the southern Altiplano – with few economic resources and marginalised by the state – were able to mobilise their local, well organised communities to generate opportunity. As Burnett and Murphy indicate, 'while imperfect, fair trade does embody elements of a Polanyian Double Movement, that is, a social movement that emerges in confrontation of existing economic structures ... with an effort to re-embed markets in society' and it 'also provide[s] important opportunities for farmers, most of whom have too few alternatives and [is] evidence that not all small-scale producers are pursuing the same model of governance'.[44]

With little hope of accessing domestic markets for their products, quinoa producers forged long-distance trade relationships – a pre-colonial strategy that not only ensured their survival, but spawned a socioeconomic revival. Nonetheless, new and profound challenges to sustainable 'quinoa re-peasantisation' have also emerged.

Challenges to sustainable re-peasantisation in the southern Altiplano

Transformation of land and resource use

The southern Altiplano is the fastest expanding region of quinoa cultivation in Bolivia. Already high producer prices for quinoa relative to other smallholder crops skyrocketed in 2008, more than tripling between 2008 and 2010. Recent reports claim that prices doubled in 2013 alone, a fact largely attributed to publicity from the UN's International Year of Quinoa.[45] This spike has promoted the expansion of the agricultural frontier, more than doubling the area planted in the Altiplano in four years – from 51,000 hectares in 2009 to an estimated 104,000 hectares in 2013.[46] This expansion poses a potential threat to the fragile, sandy and volcanic soils of the region, which are characterised by high salinity, a scarcity of organic matter and low moisture retention capacity.[47]

While the hillsides contain higher amounts of clay, organic matter and nutrients than the flatlands, many hillside plots are now abandoned, as farmers prefer to cultivate the pampas with tractors. By loosening the subsoil, the use of disk ploughs and sowing machinery has created a more favourable environment for pests.[48] Additionally, fallow periods of six to eight years have given way, in some areas, to near continuous production.[49]

Until the introduction of tractors for agricultural production in the 1970s, pastoralism had been the primary economic activity of the southern Altiplano, providing critical fertility for subsistence quinoa plots. Indeed, the relationship between quinoa, llamas and humans represents an ancient and pervasive form of symbiosis.[50] Higher prices in the 1980s, however, motivated families with larger herds to sell their llamas or sheep in order to invest in machinery and expand quinoa production on communal grazing lands.[51] A shortage of labour as a result of out-migration also stimulated the shift away from animal husbandry, which requires daily care and is ultimately less remunerative.[52]

The reduced area and labour time devoted to pastoralism has begun to generate a rupture in the 'quinoa–camelid complex' which has been acutely felt, for example, in the high cost of animal manure. Quinoa producer and ANAPQUI member Daniel comments:

> Before, my grandparents always had manure, from sheep and llamas. Not many people had pick-up trucks back then – just a few people. When they came, my grandmother would give it away for free. But today, a truckload can cost you 2000 to 3000 bolivianos [$385–430].[53]

The value of animal manure, meat, and fibre, however, has thus far not made pastoralism profitable enough – considering its high labour costs – to help it compete with quinoa and recover the ecological balance between crops and animals.

While a broad-based extension programme to support sustainable quinoa production throughout the sector is lacking, a number of localised, mostly peasant-led initiatives exist. ANAPQUI, for instance, provides assistance for sustainable production through its technical arm PROQUINAT (Program for the Production of Natural Quinoa/*Programa de Producción de Quinua Natural*). In 2010 ANAPQUI also formed its own financial entity, the Agro-livestock Financial Association of the Southern Altiplano (*Financiera Asociación Agropecuaria del Altiplano Sur*, FAAAS), which provides credit for llama production as part of an effort to promote soil fertility through llama–quinoa integration.[54]

Returning migrants and conflicting rationalities

The quinoa sector is often hailed for its contribution to the repopulation of a region previously hollowed out by out-migration, infusing new life into the countryside.[55] Many comment that previous waves of out-migration had left the region inhabited primarily by elderly people, and lacking the resources and labour to invest in the communities. Yet returning migration has also amplified local tensions, as described below.

According to the most recent national census (2012), the country's eight biggest quinoa-producing provinces have registered an average annual population growth rate of 19.25% since the last census was taken in 2001.[56] Within these provinces the quinuero municipalities of Nor Lípez, Sur Lípez and Salinas de Garci Mendoza registered growth rates as high as 34.4%, 39.3% and 25.5%, respectively. These astronomical rural population growth rates compare to an almost stagnant national average rural population growth rate of 0.5% and a national urban population growth rate of 2.4%.[57]

For returning migrants the experience can be quite emotionally charged, as Gustavo describes below. Gustavo has lived his entire life in the capital La Paz, but five years ago began travelling to the southern Altiplano to tend his quinoa field in his father's native village. However, he was not welcomed with open arms when he first arrived. Not knowing exactly where his family's land was located, he found it difficult to get answers from community members. Despite the initially icy reception, Gustavo was profoundly touched by the reconnection to his rural roots:

> I'm returning now to my ancestors' land. My father had left the village in the fifties. He always stayed in touch with his roots though, even though he didn't produce much, just enough for the family. Now, with quinoa I'm going back. When I got there, it was like finding myself. This is my land. This is where I come from.[58]

Gustavo and his father are characteristic of a common Andean phenomenon of double or even triple residency. Those who have left their native communities – but who have not abandoned their lands – are paradoxically known as 'residents' (*residentes*). This generally refers to the fact that they have become *urban* residents who no longer live in the countryside.[59] Double residency represents a kind of risk-aversion that allows for the possibility of returning to subsistence farming if needed. Abandoning or selling one's land is an act of great finality that is not done without a secure economic alternative or access to land elsewhere.[60]

Those who remain in the community, by contrast, are known as *estantes*. With regards to quinoa production, there is an apparent clash of rationalities between 'those who stayed' and 'those who left' (and have recently returned); in other words, those who *live* in the community (*estantes*) as opposed to those who only *farm* in the community (*residentes*). First, there is a perceived divide related to the *residente*'s generally higher level of formal education and link to urban-based power structures (political parties, government posts). According to *estante* Efraín, 'the new generation are professors and professionals who don't respect the elders who can hardly read and write; because of this, it's been difficult to make them fulfil their duties'.[61]

Second, 'returning' migrants are often seen as having neglected their responsibilities – such as road maintenance or taking on rotating leadership posts – while they were away. Third, many *residentes* manage their production remotely, neglecting long-standing community norms about land and resource use, eg norms regulating fallow periods and crop rotations, in order to produce more quinoa, leading to intra-community and even intra-familial resource conflicts.[62]

When asked how higher quinoa prices have changed community life, Pedro, a quinoa producer and *estante*, gives a complex answer that points to the tension between *estantes* and *residentes*:

> Quinoa has improved our quality of life. Before, when the price was low, people left, migrated to the cities, they became *residentes* and we barely saw them anymore. But with the increase in prices, those people have returned – but as strangers.

[Has this been positive for the communities?]

> No, it's been negative, because they just came back for the price. They come to plant, and then they come to harvest, but the rest of the year they're nowhere to be found. Some even come to harvest too late, when the quinoa is already drying out in the fields and going bad. People here have their beliefs, you know? Sometimes people say, 'They're making the quinoa suffer! Because of this, it won't rain this year. Things are going to go poorly for us because of the *residentes*'.[63]

These dynamics demonstrate that sustainability is not merely a technical question. It is tightly linked – as it has been for millennia – to culturally embedded organisational forms that mediate resource use and land tenure. Having survived for centuries on the margins of colonial and postcolonial development, the *ayllu* now faces profound challenges.

Territory and land control

Recent literature on the new 'land grabs' has pointed out that a focus on foreign land acquisitions and 'mega-deals' tends to miss or underplay the role of local government in facilitating land grabbing; deals led by domestic capital; and smaller-scale land acquisitions.[64] Hall's comparative work on crop booms in Southeast Asia further complicates the narrative of large scale, foreign-led land grabs, indicating that, under 'boom' conditions, not just domestic capital but also smallholders themselves may become agents of land grabbing.[65]

While the issue in the southern Altiplano has not been outsider 'land grabs', changing mechanisms of land control have allowed individuals with membership ties to indigenous communities to expand private, individualised production on communal lands. The National Coordinator of Agronomists and Veterinarians Without Borders (AVSF) in Bolivia, explains:

> Prior to mechanisation, the criteria for determining a family's access to land corresponded to a family's size and capacity – in other words, the number of bodies it had [available to work] and mouths it had to feed. So the community would allocate a parcel, the size of which varied in direct proportion to the number of family members and their subsistence needs. Now, the big shift is that it's the amount of capital the family has that determines how much land it can control, because capital means the ability to invest in mechanisation. So with a tractor you can cover quite a bit of land, maybe 40 or 50 hectares or even more. So there's a bit of a

spiral that makes the community controls break down, especially the ancestral norms that once regulated access to land.[66]

In much of the region land is not owned as private property, but rather held as communal indigenous territories under a communal title known as a 'communal territory of origin' (*tierra comunitaria de origen*, TCO). TCOs are a form of communal title hard-won by indigenous social movements in the 1980s and 1990s and institutionalised by the 1994 land reform law (*Ley INRA*). In theory, TCOs should protect indigenous lands from outside profiteers and market forces. To some extent they have, by placing indigenous territories outside the land market. But the assumption that all members of a TCO necessarily operate in a way conducive to the conservation of natural resources and local culture is difficult to maintain.

Richard, an *estante* and quinoa farmer, points to the complex overlap of community governance (*ayllus*), collective land titles (TCOs) and the aspirations of individuals (*estantes* and *residentes*) in his community:

It's prohibited to buy and sell land because these are communal lands, a TCO. So nobody is the owner. The community decides how it should be managed, how much of it should be under production. It's prohibited to cede your land to anyone from outside the community.

[Are there people from *within* the community who have expanded their production on communal lands?]

Yes. Many people became interested in quinoa before we [the *estantes*] did. I was living here permanently, but I didn't get interested in quinoa right away. Other people saw the opportunity and came back here to begin growing it. We were more concerned with stability. We weren't very ambitious. But other members of the community had this vision of growing rapidly, of having lots of cars [laughs] ... The rest of us thought it was more important to take care of the earth, to leave a legacy to our children, so that they will be able to enjoy this land.[67]

Richard's observations point to changing patterns of land control, even within the restrictions of the *ayllu* and TCO, as community members (both *residentes* and *estantes*) are able to appropriate communal lands for personal gain. While Richard now grows quinoa for sale to a private company, his comments also highlight an ongoing rift between *estantes* and *residentes* over the conflicting goals of 'stability' and sustainability over time, on the one hand, versus accumulation, on the other.

Despite increased opportunities to live from agriculture in the southern Altiplano thanks to quinoa markets, multiple tensions exist, one of which is between the community-based logic of *estantes* and the seemingly more extractivist logic of *residentes*.[68] There is also an increasingly individualised notion of land use, provoked in part by mechanisation, which is no longer as responsive to communal norms governing sustainable practices. This serves to undermine indigenous resource management, suggesting that 'quinoa re-peasantisation' as it now stands may not be a sustainable phenomenon.

Contested re-peasantisation and 'living well'

Many producers, communities and organisations in the southern Altiplano are keenly aware of the transformations afoot in the quinoa sector and of the potential threats to sustainability and social cohesion that the boom represents. This section looks at some of the ways in which peasants struggle to harness the quinoa boom as a force of *sustainable* re-peasantisation grounded in ancestral norms, sustainable practices and local definitions of 'living well'.

For Ormachea and Ramírez, 'the return of residentes to grow quinoa in no way suggests the recreation of a "peasant" society in these communities nor a process of "repeasantization"'.[69] For these authors what is occurring in the southern Altiplano is a classic example of the advance of agrarian capitalism and Leninist differentiation. Bebbington, however, cautions against such linear and fatalistic predictions, suggesting instead that Andean peasants have time and again demonstrated the ability to 'expand their control over livelihood and landscape change and so negotiate globalization processes'.[70]

Walsh-Dilley, for example, argues that peasants in San Juan, Potosí, are so firmly rooted in a 'moral economy' that they are able to engage with the quinoa market 'as an opportunity rather than a compulsion'.[71] She argues that reciprocity relations in this community have actually been strengthened, not weakened, as peasants increasingly make use of non-market and cooperative mechanisms to access scarce labour resources and expand quinoa production for global markets.[72]

Clearly there is great diversity both among and within communities of the region as to the degree and character of the quinoa boom. Factors that might affect its impact include topography; distance from markets including labour markets; distance from and quality of roads; and the presence or absence of committed, forward-thinking community leaders. Nonetheless, Walsh-Dilley's community ethnography indicates that we would do well to heed Bebbington's advice to:

> [employ] caution before uncritical acceptance either of the empirical assertions or of the normative tone of crisis narratives on the demise of rural livelihoods, the destruction of rural environments and the disempowerment of rural communities in the face of global integration. These may well be frequent outcomes, but not inevitable ones.[73]

A number of communities, for instance, have initiated community meetings or workshops that bring together both *residentes* and *estantes* to discuss quinoa cultivation (among other matters) and to attempt to devise ecologically and culturally appropriate solutions. Part of this work has involved collectively remembering, recovering and redefining ancestral norms and land-use practices such as the traditional system of sectoral fallowing known as *mantos*.

Community norms often go beyond land and resource use, requiring producers to become active in the communities, to attend community meetings, to help solve communal problems, and to invest their profits in the community's wellbeing. Walter Mamani, a quinoa producer and faculty member at the Technical University of Oruro (UTO), explains:

In some communal norms, they outlined that the producer who wishes to grow quinoa has to build a house in the community ... Some communal norms have outlined that families have to invest in improving their kitchens. Before, when there wasn't much money, people would cook with dirty water, or kids would get sick because they didn't have warm coats. These things can be addressed through the communal norms, because now there's economic growth.[74]

At the heart of these efforts is the desire, not to recreate some idyllic version of the past, but to reassert collective decision making over individual accumulation so that quinoa cultivation may contribute to common well-being or 'living well' – known as *sumaq qamaña* in the Aymara language; *sumak kawsay* in quechua; and *buen vivir* or *vivir bien* in Spanish – for generations to come.

In recent years the notion of living well has been embraced by various indigenous movements throughout Latin America; incorporated into the new constitutions of both Ecuador and Bolivia; and adopted by the global movement for climate justice.[75] It expresses, on one hand, 'critical reactions to classical Western development theory [and] on the other hand, it refers to alternatives to development emerging from indigenous traditions, and in this sense the concept explores possibilities beyond the modern Eurocentric tradition'.[76]

In a survey conducted by UTO, families in 18 communities of the municipality of Salinas, in the heart of quinoa country, were asked to define 'living well'. Above all other definitions the families of Salinas defined living well as 'living in harmony', ie without social conflicts within or among families and communities. For Mamani, 'In an area where the quinoa boom has created this issue of land conflicts, it's significant that living well for them means living in harmony'.[77] Other aspects of living well that were mentioned were: a dignified home; a healthy diet; access to education; maintaining cultural identity; and conserving natural resources so that they may benefit future generations.[78]

According to the Bolivian scholar and agronomist Mario Torrez, '*suma qamaña* operates in a special social, environmental, and territorial context, represented by the Andean *ayllu* ... It is a space of well-being with people, animals, and crops [in which] there is no duality that separates society from Nature since one contains the other and they are inseparable complementarities'.[79] Of course, 'living well' is subject to numerous contradictions, appropriations and distortions, especially when deployed by the state.

Yet perhaps at its best – defined and defended by local populations – living well's 'various expressions, whether old or new, original or the product of different hybridizations, open the door to another path'.[80] This 'other path' is as yet unclear in the quinoa sector; it is being fashioned and debated by individuals, communities and organisations. At stake is a contested process of re-peasantisation whose character and sustainability have yet to be seen.

Conclusion

The southern Altiplano, I argue, occupies what Van der Ploeg calls a 'grey zone' at the interface between 'peasantness' and entrepreneurial farming. In this grey zone some non-peasants are returning to the countryside to farm; some peasants are constituting themselves as entrepreneurs; and other peasants are working to reshape their social and productive system so as to protect and enhance local culture, autonomy and natural resources. As Van der Ploeg notes,

> In these grey zones one encounters *degrees of peasantness* that are far from being theoretically irrelevant. Indeed they characterize arenas in which, over time, important fluctuations occur with respect to de-peasantization and re-peasantization...Both processes will pass through many in-between situations, thus enlarging the many shades of grey that together characterize this intersection.[81]

As Desmarais points out in her study of La Vía Campesina, 'communities should be seen as sites of diversity, differences, conflicts, and divisions often expressed along gender, class, and ethnic lines and characterized by competing claims and interests'.[82]

Despite the highly fraught transformations occurring in the southern Altiplano, there are promising grassroots organising efforts, both at the level of producers' associations and at the level of indigenous *ayllus*, *markas* (a grouping of *ayllus*) and confederations. CONAMAQ (the National Council of Ayllus and Markas of Qullasuyu)[83] has been calling for the government to prioritise domestic consumption of quinoa as a means of strengthening cultural identity and tackling malnutrition.[84] ANAPQUI is working with members through its technical arm PROQUNAT to promote agro-ecology and a culturally appropriate development model. The renewed invocation of ancestral land-use norms, the creation of new local rules to regulate how wealth is invested in community, and reflections on what it means to 'live well' are all examples of a process of re-peasantisation that is both contested and under construction.

If the *state* has been conspicuously absent from this analysis, it is because, as outlined in the historical section of this article, the modern state – from the colonial era to the present day – has itself been conspicuously absent from the southern Altiplano region. While the current government has discursively attempted to take some credit for the quinoa boom – for instance through successful lobbying at the UN level for an 'International Year of Quinoa' (2013) – state interventions in the sector have been negligible. As ANAPQUI Marketing Director Juan Carlos observes, 'a lot of people think that the government created the quinoa boom, but that's not the case; it's the producers, along with our clients and the consumers, who [made it happen]'.[85] With apparently little political will to enact state extension programmes or regulate supply – at least for the time being – peasant, grassroots efforts are leading the charge in trying to secure their own, sustainable livelihoods in the southern Altiplano.

This article has argued that the recent 'quinoa boom' in Bolivia has its roots in a decades-long process of re-peasantisation in which indigenous peasants have struggled – collectively and individually – to defend, rediscover and redefine their 'degree of peasantness', while navigating neoliberalism and global market forces. Re-peasantisation in this case did not occur through a retreat from the market or return to the local, but rather by leveraging collective indigenous organisation in order to forge global relationships and access export markets. Through the grassroots efforts of peasant organisations and other civil society actors, a market for this maligned indigenous crop was generated against sharp odds, at the height of neoliberalism.

While there is little doubt that the quinoa export sector has benefited communities in the region, it has also created steep new challenges, not least of which is the influx of returning migrants with distinct, and even conflicting, rationalities. The strong history of resistance and autonomy in the southern Altiplano, however, may bode well for the region's ability to assert an alternative model of production that sustains communities; but this remains to be seen.

For the food sovereignty movement this case shines light on the need to examine the culturally and historically specific challenges facing re-peasantisation in particular places. It should not be assumed, for instance, that processes of re-peasantisation only or necessarily occur via a retreat from (global) markets and a return to subsistence. In the case of quinoa, engagement with alternative global food networks has allowed peasant associations of the southern Altiplano to build a successful economy in a region marked by poverty and out-migration. This complex process of re-peasantisation, however, has generated both opportunities and challenges. It is equally important for advocates of food sovereignty not to simply celebrate re-peasantisation while neglecting how fraught such processes can often be.

If re-peasantisation is to be seen as an essential process for building food sovereignty, then the many challenges of reconstructing peasantries must be acknowledged and explored. Admittedly, the biggest challenges to the world's diverse peasantries may stem from the expansion and restructuring of the capitalist food system, including the financialisation of agriculture and corporate land grabs. But they also include numerous *internal* tensions, conflicts and contradictions occurring within those 'peasantries under [re-]construction'. How these tensions are reconciled will doubtless determine their – and our – ability to resist the corporate onslaught and to build food sovereignty, 'living well' and other ecologically and culturally appropriate alternatives.

Acknowledgements

I am deeply grateful to the dozens of individuals and organisations in Bolivia and the USA who lent their generous collaboration to this research. I also wish to thank Christina Bronsing-Lazalde, Carly Finkle and Brock Hicks for their invaluable research assistance, and the editors and anonymous reviewers of this special issue for their thoughtful feedback. As always, any errors or flaws in the analysis are my own.

Notes

1. Quinoa is not a cereal grain, but rather a cereal-like seed, which is why it is often referred to as a 'pseudocereal.'

2. Interviews fell into three subject groups – quinoa producers, private sector actors and 'experts' (academics and NGO workers) – and were designed to ascertain subjects' understanding and interpretation of the opportunities and challenges presented by the recent boom in Northern consumer demand for quinoa. The names of all interview subjects have been changed, and identifying information removed, to protect their confidentiality.
3. Desmarais et al., *Food Sovereignty*, 2.
4. Desmarais, *La Vía Campesina*, 37.
5. World Health Organization (WHO), "Urban Population Growth."
6. Araghi, "Global Depeasantization," 338.
7. Kay, "Reflections on Latin American Rural Studies," 926.
8. Radcliffe, "The Geographies of Indigenous Self-representation in Ecuador," 16.
9. Bernstein, *Class Dynamics of Agrarian Change*.
10. Van der Ploeg, "The Peasantries of the Twenty-first Century," 21.
11. Van der Ploeg, *The New Peasantries*, 7.
12. Ibid, emphasis added.
13. Burnett and Murphy, "What Place for International Trade in Food Sovereignty?," 4.
14. Bernstein, "Food Sovereignty via the 'Peasant Way'," 1032 (emphasis added).
15. Ibid., 1046.
16. Van der Ploeg, "Peasant-driven Agricultural Growth," 1022.
17. Bebbington and Batterbury, "Transnational Livelihoods and Landscapes," 370.
18. Tapia, *Cultivos Andinos Subexplotados*.
19. Murra, "The Economic Organization of the Inca State."
20. Hellin and Higman, "Crop Diversity and Livelihood Security in the Andes."
21. D'Altroy, "Andean Land Use at the Cusp of History"; and Kolata, *Ancient Inca*.
22. Dong et al., "Vulnerability of Worldwide Pastoralism to Global Changes," 9.
23. Kolata, *Ancient Inca*.
24. Healy, "Towards an Andean Rural Development Paradigm?," 28.
25. Healy, *Llamas, Weavings, and Organic Chocolate*; and Brett, "The Political-Economics of Developing Markets."
26. Pérez et al., "The Promise and the Perils of Agricultural Trade Liberalization."
27. Ibid.
28. Cazorla et al., *Rural Migration in Bolivia*.
29. Ibid; and Pérez et al., "The Promise and the Perils of Agricultural Trade Liberalization."
30. Desmarais et al., *Food Sovereignty*.
31. Goodman et al., *Alternative Food Networks*.
32. Laguna, "El Impacto del Desarrollo del Mercado de la Quinua."
33. Healy, *Llamas, Weavings, and Organic Chocolate*.
34. Rojas et al., *Granos Andinos*. Rojas et al. identify five quinoa 'ecotypes' associated with different Andean regions: sea level (primarily coastal Chile); yungas (1500–2000 m); valleys (2500–3500 m); northern and central Altiplano (Peru and Bolivia, where the highest diversity is found); and salt flat quinoa or 'royal quinoa' (*quinua real*) of the southern Altiplano of Boliva.
35. Laguna, "El Impacto del Desarrollo del Mercado de la Quinua."
36. Thrupp, *Bittersweet Harvests for Global Supermarkets*.
37. Ibid.
38. Interview with Juan Carlos, Marketing Director at ANAPQUI, La Paz, July 22, 2013 (author's translation).
39. Laguna et al., "Del Altiplano Sur Bolivariano hasta el Mercado Global," 68 (author's translation).
40. AVSF, *Quinua y Territorio Nuevos Desafíos*.
41. "El Consumo de Quinua en el País se Triplicó en los Últimos 4 Años." *La Razón*, February 17, 2013. http://www.la-razon.com/economia/consumo-quinua-triplico-ultimos-anos_0_1780622010.html.
42. FAOstat, "Food Balance Sheet: Bolivia (Plurinational State of)." Accessed March 28, 2014. http://faostat3.fao.org/faostat-gateway/go/to/download/FB/FB/E.
43. Perez et al., "Food Crisis, Small-scale Farmers, and Markets in the Andes."
44. Burnett and Murphy, "What Place for International Trade in Food Sovereignty?," 8.
45. "Quinua Duplica Precio en Año Internacional." Associated Press, March 21, 2014. http://noticias.latino.msn.com/latinoamerica/bolivia/quinua-duplica-precio-en-a%C3%B1o-internacional-1.
46. Fundación Milenio, "Quinua en Bolivia."
47. Vallejos Mamani et al., *Medio Ambiente y Producción de Quinoa*.
48. Jacobsen, "The Situation for Quinoa."
49. Rojas et al., *Study on the Social, Environmental and Economic Impacts of Quinoa Promotion*.
50. Kolata, "Quinoa"; and Kolata, *Ancient Inca*.
51. Laguna, "El Impacto del Desarrollo del Mercado de la Quinua."
52. Ibid.
53. Interview with Daniel, quinoa producer and ANAPQUI member, Salinas de Garci Mendoza, July 18, 2013 (author's translation).

54. AVSF, *Quinua y Territorio Nuevos Desafíos*.
55. "'Boom' de la Quinua Provoca Regreso de Migrantes." *El Potosí*, September 12, 2012.
56. INE, "67% de la Población del País Habita en Áreas Urbanas."
57. Ibid.
58. Interview with Gustavo, independent quinoa producer. La Paz, June 10, 2013 (author's translation).
59. Urioste, *Los Nietos de la Reforma Agraria*. Those who have moved to another rural region – generally migrating from the highlands to more tropical elevations – are not referred to as *residentes*, but rather as *colonizadores* or settlers.
60. Ibid.
61. Quoted in AVSF, *Quinua y Territorio*, 49, author's translation.
62. AVSF, *Quinua y Territorio*; AVSF, *Quinua y Territorio Nuevos Desafíos*; and Ormachea and Ramírez, *Propiedad Colectiva de la Tierra*.
63. Interview with Pedro, quinoa producer and COPROQUIR member (regional chapter of ANAPQUI), Irpani, July 19, 2013 (author's translation).
64. Borras et al., "Land Grabbing in Latin America and the Caribbean."
65. Hall, "Land Grabs, Land Control, and Southeast Asian Crop Booms."
66. Interview with the National Coordinator of Agronomists and Veterinarians Without Borders, La Paz, May 27, 2013 (author's translation).
67. Interview with Richard, independent quinoa producer. La Paz, June 10, 2013 (author's translation).
68. It should be noted that this dichotomy does not always hold. There are most certainly some extractivist *estantes* as well as ecologically and culturally sensitive *residentes*.
69. Ormachea and Ramírez, *Propiedad Colectiva de la Tierra*.
70. Bebbington, "Globalized Andes?," 371.
71. Walsh-Dilley, "Negotiating Hybridity in Highland Bolivia," 19.
72. Her study nonetheless identifies producers – apparently in the minority – who see reciprocal labour as a waste of time, preferring instead to employ machinery and paid workers.
73. Bebbington, "Globalized Andes?," 431.
74. Interview with Walter Mamani, quinoa producer and faculty at the Technical University of Oruro (UTO), Oruro, July 17, 2013 (author's translation).
75. Zimmerer, "Environmental Governance through 'Speaking like an Indigenous State'."
76. Gudynas, "Buen Vivir," 441.
77. Interview with Walter Mamani, quinoa producer and faculty at the UTO, Oruro, July 17, 2013 (author's translation).
78. Ibid.
79. Quoted in Gudynas, "Good Life."
80. Ibid.
81. Van der Ploeg, *The New Peasantries*, 37–38 (emphasis added).
82. Desmarais, *La Vía Campesina*, 37.
83. Qullasuyu refers to the region of the Inca empire that is present-day Bolivia.
84. "Según el CONAMAQ, el Gobierno debe Priorizar Consumo Interno de Quinua." *El Potosí*, March 18, 2013. http://www.elpotosi.net/2013/03/18/19.php.
85. Interview with Juan Carlos, Marketing Director at ANAPQUI, La Paz, July 22, 2013 (author's translation).

Bibliography

Araghi, Farshad. "Global Depeasantization, 1945–1990." *Sociological Quarterly* 36, no. 2 (1995): 337–368.
Agronomes et Vétérinaires Sans Frontières (AVSF). *Quinua y Territorio*. La Paz: Plural editores, 2009.
AVSF. *Quinua y Territorio Nuevos Desafíos: Gobernanza Local y Producción Sostenible de la Quinua Real en Bolivia*. La Paz: AVSF, February 2014.
Bebbington, Anthony. "Globalized Andes? Livelihoods, Landscapes and Development." *Cultural Geographies* 8, no. 4 (2001): 414–436.
Bebbington, A. J., and S. P. J. Batterbury. "Transnational Livelihoods and Landscapes: Political Ecologies of Globalization." *Cultural Geographies* 8, no. 369 (2001): 369–380.
Bernstein, Henry. "Food Sovereignty via the 'Peasant Way': A Sceptical View." *Journal of Peasant Studies* 41, no. 6 (2014): 1031–1063.
Bernstein, Henry. *Class Dynamics of Agrarian Change*. Halifax: Fernwood Publishing, 2010.
Borras, Saturnino M., Jennifer Franco, Sergio Gómez, Cristóbal Kay, and Max Spoor. "Land Grabbing in Latin America and the Caribbean." *Journal of Peasant Studies* 39, nos. 3–4 (2012): 845–872.
Brett, John A. "The Political-Economics of Developing Markets versus Satisfying Food Needs." *Food and Foodways* 18, nos. 1–2 (2010): 28–42. doi:10.1080/07409711003708249.
Burnett, Kim, and Sophia Murphy. "What Place for International Trade in Food Sovereignty?" *Journal of Peasant Studies*, 2014. doi: 10.1080/03066150.2013.876995.

FOOD SOVEREIGNTY

Cazorla, Iván, Nico Tassi, Ana Rubena Miranda, Lucia Aramayo Canedo, and Carlos Balderrama Mariscal. *Rural Migration in Bolivia: The Impact of Climate Change, Economic Crisis and State Policy*. London: IIED, 2011.

D'Altroy, Terence N. "Andean Land Use at the Cusp of History." In *Imperfect Balance: Landscape Transformations in the Pre-Columbian Americas*, 357–390. New York: Columbia University Press, 2000.

Desmarais, Annette Aurelie. *La Vía Campesina: Globalization and the Power of Peasants*. Ann Arbor, MI: Pluto Press, 2007.

Desmarais, Annette, Nettie Wiebe, and Hannah Wittman, eds. *Food Sovereignty: Reconnecting Food, Nature and Community*. Halifax: Fernwood Publishing/Food First Books, 2010.

Dong, Shikui, Lu Wen, Shilang Liu, Xiangfeng Zhang, James P. Lassoie, Shaoliang Yi, Xiaoyang Li, Jinpeng Li, and Yuanyuan Li. "Vulnerability of Worldwide Pastoralism to Global Changes and Interdisciplinary Strategies for Sustainable Pastoralism." *Ecology and Society* 16, no. 2 (2011) [online]. http://www.ecologyandsociety.org/vol16/iss2/art10/.

Fundación Milenio. "Quinua en Bolivia: Fortalezas y Debilidades." *Informe Nacional de Coyuntura* 190, April 19, 2013.

Goodman, David, Melanie E. DuPuis, and Michael K. Goodman. *Alternative Food Networks: Knowledge, Practice and Politics*. London: Routledge, 2012.

Gudynas, Eduardo. "Buen Vivir: Today's Tomorrow." *Development* 54, no. 4 (2011): 441–447.

Gudynas, Eduardo. "Good Life: Germinating Alternatives to Development." *América Latina en Movimiento*, July 14, 2011. http://alainet.org/active/48054.

Hall, Derek. "Land Grabs, Land Control, and Southeast Asian Crop Booms." *Journal of Peasant Studies* 38, no. 4 (2011): 837–857.

Healy, Kevin. *Llamas, Weavings, and Organic Chocolate: Multicultural Grassroots Development in the Andes and Amazon of Bolivia*. Notre Dame, IN: University of Notre Dame Press, 2001.

Healy, Kevin. "Towards an Andean Rural Development Paradigm?" *NACLA Report on the Americas* 38 (2004): 28–33.

Hellin, Jon, and Sophie Higman. "Crop Diversity and Livelihood Security in the Andes." *Development in Practice* 15, no. 2 (2005): 165–174. doi:10.1080/09614520500041344.

Instituto Nacional de Estadísticas (INE). "67% de la Población del País Habita en Áreas Urbanas y 32.7% en Áreas Rurales." January 6, 2014. http://www.censosbolivia.bo/sites/default/files/archivos_adjuntos/N%204%20Area%20urbanas%20y%20rurales_1.pdf.

Jacobsen, S.-E. "The Situation for Quinoa and its Production in Southern Bolivia: From Economic Success to Environmental Disaster." *Journal of Agronomy and Crop Science* 197, no. 5 (2011): 390–399. doi:10.1111/j.1439-037X.2011.00475.x.

Kay, Cristóbal. "Reflections on Latin American Rural Studies in the Neoliberal Globalization Period: A New Rurality?" *Development and Change* 39, no. 6 (2008): 915–943.

Kolata, Alan L. *Ancient Inca*. New York: Cambridge University Press, 2013.

Kolata, Alan L. "Quinoa: Production, Consumption and Social Value in Historical Context." University of Chicago Department of Anthropology, 2009. http://lasa.international.pitt.edu/members/congress-papers/lasa2009/files/KolataAlanL.pdf.

Laguna, Pablo. "El Impacto del Desarrollo del Mercado de la Quinua en los Sistemas Productivos y Modos de Vida del Altiplano Sur Boliviano." In *XVI Simposio de La Asociación Internacional de Sistemas de Producción*. Santiago, 2000. http://www.academia.edu/2190139/El_impacto_del_desarrollo_del_mercado_de_la_quinua_en_los_sistemas_productivos_y_modos_de_vida_del_Altiplano_Sur_boliviano.

Laguna, Pablo, Aurélie Carmentrand, and Zina Cáceres. "Del Altiplano Sur Bolivariano hasta el Mercado Global: Coordinación y Estructuras de Gobernancia de la Cadena de Valor de la Quinua Orgánica y del Comercio Justo." *Agroalimentaria* 11, no. 22 (June 2006).

Murra, John V. "The Economic Organization of the Inca State." PhD Diss., University of Chicago, 1956.

Ormachea, S. Enrique, and Nilton Ramírez F. *Propiedad Colectiva de la Tierra y Producción Capitalista: El Caso de la Quinua en el Altiplano Sur de Bolivia*. La Paz: CEDLA, 2013.

Perez, Carlos A., Claire Nicklin, and Sarela Paz. "Food Crisis, Small-scale Farmers, and Markets in the Andes." *Development in Practice* 21, nos. 4–5 (2011): 566–577. doi:10.1080/09614524.2011.562486.

Pérez, Mamerto, Sergio Schlesinger, and Timothy A. Wise. "The Promise and the Perils of Agricultural Trade Liberalization: Lessons from Latin America." Washington Office on Latin America (WOLA), June 2008. http://ase.tufts.edu/gdae/Pubs/rp/AgricWGReportJuly08.pdf.

Radcliffe, Sarah A. "The Geographies of Indigenous Self-representation in Ecuador: Hybridity and Resistance." *Revista Europea de Estudios Latinoamericanos y del Caribe* 63 (December 1997): 9–27.

Rojas, Wilfredo, José Luis Soto, and Enrique Carrasco. *Study on the Social, Environmental and Economic Impacts of Quinoa Promotion in Bolivia*. La Paz: Fundación PROINPA, 2004. http://www.cropsforthefuture.org/publication/strategic-document/Study%20on%20the%20social,%20environmental%20and%20economic%20impacts%20of%20quinoa%20promotian%20in%20Bolivia.pdf.

Rojas, Wilfredo, José Luis Soto, Milton Pinto, Matthias Jager, and Stefano Padulosi. *Granos Andinos: Avances, Logros y Experiencias Desarrolladas en Quinua, Cañahua y Amaranto en Bolivia*. Biodiversity International, 2010. http://www.proinpa.org/tic/pdf/Quinua/Varios%20quinua/pdf35.pdf.

Tapia, Mario E. *Cultivos Andinos Subexplotados y su Aporte a la Alimentación*. Rome: Food and Agriculture Organization, 1990.

Thrupp, Lori Ann. *Bittersweet Harvests for Global Supermarkets: Challenges in Latin America's Agricultural Export Boom*. Washington, DC: World Resources Institute, 1995.

Urioste F. de C., Miguel. *Los Nietos de la Reforma Agraria: Acceso, Tenencia y Uso de la Tierra en el Altiplano de Bolivia*. La Paz: Fundación Tierra, 2005.

Vallejos Mamani, Pedro Román, Zaima Navarro Fuentes, and Dante Ayaviri Nina. *Medio Ambiente y Producción de Quinua: Estrategias de Adaptación al Cambio Climático*. La Paz: Programa de Investigación Estratégica en Bolivia (Fundación PIEB), 2011.

Van der Ploeg, Jan Douwe. "Peasant-driven Agricultural Growth and Food Sovereignty." *Journal of Peasant Studies* 41, no. 6 (2014): 999–1030.

Van der Ploeg, Jan Douwe. "The Peasantries of the Twenty-first Century: The Commoditisation Debate Revisited." *Journal of Peasant Studies* 37, no. 1 (2010): 1–30.

Van der Ploeg, Jan Douwe. *The New Peasantries: Struggles for Autonomy and Sustainability in an Era of Empire and Globalization*. London: Earthscan, 2009.

Walsh-Dilley, Marygold. "Negotiating Hybridity in Highland Bolivia: Indigenous Moral Economy and the Expanding Market for Quinoa." *Journal of Peasant Studies* 40, no. 4 (2013): 659–682.

World Health Organization. "Urban Population Growth" *Global Health Observatory*. Accessed 15 November 2014. http://www.who.int/gho/urban_health/situation_trends/urban_population_growth_text/en/

Zimmerer, Karl S. "Environmental Governance through 'Speaking Like an Indigenous State' and Respatializing Resources: Ethical Livelihood Concepts in Bolivia as Versatility or Verisimilitude?" *Geoforum*, 2013. doi:http://dx.doi.org/10.1016/j.geoforum.2013.07.004.

Food sovereignty as praxis: rethinking the food question in Uganda

Giuliano Martiniello

Makerere Institute of Social Research (MISR), Makerere University, Kampala, Uganda

This article critically reflects upon conceptual and analytical questions that affect the practical implementation of food sovereignty in Uganda, a country often labelled as the potential breadbasket of Africa. It proposes to look at the integration of food and land-based social relations in the context of localised and historical–geographical specificities of livelihood practices among Acholi peasants in northern Uganda as a way to ground the concept. It argues that many of the organising principles at the core of the food sovereignty paradigm are inscribed in the socio-cultural and ecological practices of peasant populations in northern Uganda. Yet these practices are taking place in an increasingly adverse national and international environment, and under circumstances transmitted from the past, which enormously challenge their implementation and jeopardise the future of food security and sovereignty prospects for peasant agriculture.

Emerging debates on food sovereignty

The global food crisis of 2008 generated a throng of food riots across the global South, as consumers increasingly confronted volatile food supplies and sky-rocketing prices.[1] Although the crisis affected developing countries differentially, by the year's end the emergency had raised food import bills by an average of 37%,[2] ultimately adding another 75 million people to the ranks of the hungry and 125 million to the category of extreme poverty.[3]

Various explanations have been given for this price volatility. They include the rise of middle class demands; changing diets in countries like Brazil, India and China; and the associated growing demand for animal feed, energy security, and mitigation of climate change. Though significant, these explanations represent only partial readings of the crisis, to which we should add speculation in commodity futures (financialisation of the food sector), conversion of farmland into urban estates, and the conversion from food production to agro-fuels in many regions. A major shortcoming of these conventional explanations of the food crisis is that they are epiphenomenal in character, largely based on supply–demand analysis. In such a framework hunger and malnutrition become associated with a preoccupation with production and supply constraints, not with uneven access to productive resources. In short, these conventional explanations only highlight the symptoms rather than the underlying causes.

At root the crisis was an intrinsic manifestation of both short-term conjunctures and long-term dynamics of the neoliberal corporate-based food regime.[4] As many scholars have noted, the capitalist transformation of agriculture has involved expanding mechanisation, chemicalisation, land concentration, dependence on hydrocarbon farm inputs, expanded use of bio-technologies and commodification of seed production. Tony Weis interprets these conjunct processes as simultaneously showing the accelerating biophysical contradictions of industrial agriculture and the limitations of the emerging 'grain–oilseed–livestock complex'.[5] From a global perspective Araghi argues that this development had produced absolute de-peasantisation and displacement under postcolonial neoliberal globalism.[6]

Food sovereignty emerged in the early 1990s through La Via Campesina, a unifying collective of transnational agrarian movements. The concept represented an epistemic, theoretical and practical response to the rising consequences of the corporate food regime and the neoliberal globalisation in which it operated; a structuring that gave rise to the food crisis of 2008.[7] 'Food sovereignty' was coined with the aim of politicising the food and agricultural debates 'from below', emphasising the 'right of nations and peoples to control their own food systems, including their own markets, production modes, food cultures and environments'.[8] By questioning the assumptions that rest at the core of industrial agriculture, the concept affords a combined critique of the power, economic and ecological dynamics that characterise the international food regime.

Although containing a host of facets, food sovereignty, at its core, underscores the significance of small-scale farming and its associated agricultural practices, reasserting the vital importance of bio-diverse and sustainable agriculture.[9] Kloppenburg interprets the interlinked processes of development of transgenic varieties, bio-piracy and the imposition of intellectual property rights as moments of accumulation by dispossession.[10] Moreover, the long-term process of improving the adaptation and quality of seeds, plants and crops has traditionally been in the peasant cultural repertoire of activities. For Miguel Altieri food sovereignty is indissolubly linked to agro-ecologically-based production systems in which 'ecological interactions and synergisms between biological components provide mechanisms for the system to sponsor soil fertility, productivity and crop protection'.[11]

Within this vein food sovereignty has helped progressively advance a new set of ideas and practices related to territory, locality alternative food networks, localisation of economies and agro-ecological practices. In this regard Van der Ploeg suggests that the power of food sovereignty as a concept lies in its promise to integrate a wide range of important issues, from the quality, quantity and availability of food, to the identity of producers, style of farming, democratic representation and sustainability.[12] As Agarwal suggests, such integration has pushed food sovereignty to shift emphasis away from national self-sufficiency to local self-sufficiency and decision making autonomy[13] – a shift that has politically and intellectually resituated food sovereignty *vis-à-vis* food security.

Yet food sovereignty has had its fair share of critiques. A prominent critic of the concept suggests that it does not capture the socially differentiated nature of rural communities, such as class identities, gender, age and ethnicity.[14] Other critics have raised the question of whether or not the notion of food sovereignty has any substantive meaning,[15] and highlighted its semantic distance from rural people whose everyday existence is subordinated to a set of insecurities of land access, soil quality, water and seeds.[16] Subjected to the laws of restructuring of neoliberal capitalism, peasants, we are told, are 'disappearing',[17] or live under worsening conditions of 'commoditization of subsistence' as petty-commodity producers.[18] Furthermore, by analysing the mounting 'feminization of agriculture', Agarwal takes issue with another assumption of the food sovereignty discourse: the alleged preference of small-scale farmers for subsistence crops.[19]

The food sovereignty discourse has a particular echo in Uganda, a country often labelled as the potential breadbasket of Africa. In Uganda 75% of the country's population is engaged in agriculture-based livelihoods, and women produce some 80% of the food consumed nationally.[20] Like other developing countries, Uganda is rich in natural resources and thanks to favourable climatic conditions a great variety of food crops can be grown. Although the contribution of agriculture to total GDP has declined over the years, the sector has continued to play a vital role in the Ugandan economy, comprising 22.9% of GDP in 2011.[21] Export commodities, such as coffee, cotton, tea and tobacco, still dominate this sector, however, representing 31.4% of total export value in global value chains that stretch from East Asia and Eastern Africa to the EU.[22] And amid this commodity production within the agricultural economy, nearly 7.5 million Ugandans remain in hunger and poverty.[23] In Uganda the neoliberal discourse of food security is based on the imperative of progressively increasing the quantities of food produced through large-scale plantation systems and agribusiness technologies. At the core of the discourse is a conceptualisation and construction of food as a commodity that can be bought and sold, and whose price and quantities will be established by the law of supply and demand. This approach has shaped major policy directions aimed at land formalisation and privatisation, commercialisation of smallholder agriculture, promotion of large-scale plantations, and the introduction of genetically modified organisms (GMOs) in agriculture. Notwithstanding its primacy, this approach has been questioned by civil society organisations that contest the monopolisation of the political space by the state.

By exploring competing food-related discourses and practices, this article critically reflects upon conceptual and analytical questions that affect the practical implementation of food sovereignty in Uganda. The discussion raises the question of how the notion of food sovereignty can be operationalised in Africa, particularly in the context of the relative weakness of transnational agrarian movements, the partial co-option of civil society organisations into the political framework of 'right to food', and the predominance of neoliberal ideology surrounding food security – an ideology which lies at the core of food and agricultural modernisation policies in the country. The article emphasises the need not to analyse the food question in isolation from its constituent parts. It proposes to look at the integration of food and land-based social relations in the context of localised and historical–geographical specificities of livelihood practices among Acholi peasants in northern Uganda as a way to ground the concept. With this case study the article aims to broaden the discussion about food sovereignty by looking at it through the lenses of territory, locality, praxis and peasantries, while simultaneously providing an analysis of culture, structure and agency of local communities in the *longue durée*. Interpreting food sovereignty as praxis allows us, on the one hand, to historicise agricultural practices and strategies, and situate them within the wider contexts of social, political and economic organisation of rural households amid increased market, state policy and ecological pressures. On the other hand, it also helps facilitate discussion of agrarian conditions at the ground level from the perspective of agrarian subjects.

As I argue, the key elements in forging the social, economic and political ideals of food sovereignty in Uganda are the combined mechanisms of sovereign access to land, control over labour capacity mobilisation, and indigenous seed selection and reproduction. Moreover, I suggest that civil society organisations and peasant practices play a significant role in affecting the direction of this reform agenda, as seen in the context of increasing marginalisation, subordination and dependency in rural communities.

Pillars of food security policies in Uganda
Formalising land ownership

The idea of land formalisation as a key mechanism to increase flows of capital investment in rural areas, and thus create vibrant land markets, has rested at the core of the International Finance Institutions' (IFIs) agenda over the past decade. Of paramount relevance to this argument is the work of the Peruvian development economist, Hernando de Soto, who in the search for the 'mystery of capital' posited a direct relationship between the system of representation and formalisation of land ownership and its monetary value.[24] In his view it was this prime function of representation of land ownership, epitomised in the freehold property of the Western world, that stimulated the use of land as collateral for access to credit, and therefore as a stimulus to rural development and the commercialisation of agriculture. This discourse has been profoundly influential among development economists and policy makers, particularly in Africa. In Uganda the 1998 Land Act recognised the simultaneous presence of different forms of land tenure – freehold, leasehold, *mailo*[25] and customary. Although welcomed by land activists and NGOs, the legislative act postulated a convergence towards a homogenised, single land-tenure system based on the notion of individualised, freehold private property. The stated policy intent was to create a more harmonised and uniform land tenure system.[26] According to Mamdani, the 1998 Land Act, far from preserving and protecting customary land tenure regimes, represented the latest effort of the modern state to expand the effect and colonial logic of the 1900 Buganda Agreement to the entire country.[27]

Seen from a regional perspective, land reforms in Eastern Africa, and the changes in statutory land laws that ensued, have characterised the post-1990 period. They marked a significant continuity with colonial policies that started to promote the creation of land markets, and the demise of customary land tenure in favour of that based on individualisation.[28] This argument is resonant with another work, by Ambreena Manji, who maintains that in Africa the emphasis on neoliberal land reforms shifted the focus from land redistribution to tenure reforms, from politics to law, with the effect of depoliticising the debate and neutralising alternative paths of reform.[29] Under the advice of international organisations, foreign governments and mainstream NGOs, Uganda embraced the pro-market approach anchored in the pre-eminence of property rights. Similar to the experience of South Africa, where market-led land reforms were first promoted, this approach delivered only minimal transformation of land ownership, maintaining the status quo and preserving forms of inequality and exclusion in land-based social relations.[30]

International development agencies furthered this mono-dimensional analysis of complex land-based social relations by highlighting the importance of formal tenure security, and the associated virtues of freehold property. In this regard the recently issued World Bank survey 'Doing Better Business' exhorts the Ugandan government to increase land formally registered in the national cadastral (presently only 18%) to newly established computerised cadastral systems and digitised zonal offices.[31] Clear and secure land tenure policies are seen as the first necessary step to institute credible systems to land value, to enhance credit flows to rural areas, as well as to improve the coverage of, and level of compliance with, property taxes. In this spirit the World Bank launched the 1,000,000 land titles campaign.

The formalisation of land rights, however, has not been successful in Uganda's countryside, especially the country's northern region. There are two reasons for this: the high cost of registering land that only well-off farmers can afford; and the perceived uselessness of formalisation to smallholders. The latter contend that the traditional structures of authority and management of land in the countryside are effective in recognising the rights of members of rural communities to the use of, and residency in, land. Evidence from experiences in other countries suggests that processes of property titling and its use as collateral promote speculation, accumulation of rents and concentration of wealth via forced transfers from the less affluent to the more secure rural and urban classes.[32]

Commercialising smallholder agriculture, consolidating agribusiness

In Uganda pressure on smallholder agriculture to produce cash crops has a long history. Colonial governments used traditional chiefs to impose forced crops and forced sales.[33] The expansion of export production has consistently been the subject of moral exhortation from state leaders, missionaries and the upper class. During late colonialism and in the post-independence period 'Grow More Cotton' and 'Grow More Coffee' campaigns, under the command of local state officials, were implemented in order to expand the base of peasant production and orient it towards export.[34]

In recent times low levels of productivity of smallholder agriculture are unequivocally identified as the main cause of deteriorating agricultural production and reduced food security in the country. Over recent decades the Poverty Eradication Programs (1997–2009) and the Plan for Modernisation of Agriculture (2001–09) targeted 'model farmers' as core beneficiaries of agricultural extension services in the hope that these would act as a nucleus from which new methods of farming could spread to other village farmers. Under the slogan 'securing food security through the market' the plan aimed to increase the commercialisation of agriculture through diversification and specialisation, and new agricultural zoning. The zoning abolishing the traditional systems of agricultural extension services and marketing boards which, though focused on major export cash crops, had reduced price volatility and provided inputs and credits to farmers.[35] The stated objective was to increase farmers' income-price by reducing taxation and to establish private marketing agencies. The result was the replacement of import-substitution strategies and the further orientation of agricultural production towards export.

In the light of these market-based agricultural reforms, the government progressively reduced its public spending in agriculture from 10% in 1980 to 3.7% in 2008–09.[36] Only a tiny minority of politically well-connected individuals maintained the remaining fragments of state-sponsored agricultural support, particularly in the export sectors of coffee and other cash crops. This is illustrated by the rapid expansion of traditional export crops like coffee, cocoa, tea, sugar, cotton and palm oil. High-value horticultural commodities were also promoted, such as fresh fruits, vegetables and cut flowers, as was the expansion of large-scale production of soy, sugar and grains for biofuel and livestock feed. In 1985 food crops accounted for 72.4% of agricultural GDP. By 2000 this percentage had decreased to 65.3%.[37]

International donors and think-tanks therefore constantly exhort the government of Uganda to enable smallholder farmers to make use of inorganic fertilisers, improved seeds and planting materials, and other agricultural technologies for higher agricultural production.[38] The main objective is to ensure a higher rate of usage of appropriate fertilisers and seeds, to control crop diseases, raise the levels of capitalisation for small-scale farmers, improve rural infrastructure and connect producers to markets.[39] In line with the government's '2040 vision' of transforming Uganda from a low-income, agriculture-based country into a modern middle-income country driven by service and industry sectors, the agricultural economy needs to be transformed into a more efficient and productive sector by strengthening links between advisory services and small famers, strengthening credit institutions, prioritising road investments, and supporting the private sector to deliver agricultural services.[40]

In these discourses small-scale farming is always seen as a major part of the problem. Small-scale farming is blamed for its low levels of agricultural productivity and lack of integration within markets. In 2006, for example, the most commercialised quintile of rural households in the agricultural sector sold only about 50% of their output.[41] The solution universally advanced is the adoption of improved technologies, as prescribed by the agribusiness model, which are associated with higher agricultural incomes, improved nutritional status, lower staple food prices and increased employment opportunities.[42] In his campaign to foster agricultural commercialisation, President Yoweri Museveni, reminiscent of Green Revolution rhetoric, magnified the virtues of modern agriculture, claiming that such operations underpinned economic recovery in the country. 'Three players in agriculture have done well: the plantation owners (mostly in sugar, tea and coffee); the big scale farmers and the medium scale farmers'.[43] The political establishment in Uganda sees the rise of agricultural prices not as a threat to the social order but as an opportunity to be seized through expanding the production of export crops in which Uganda enjoys a comparative advantage. This development model promotes a larger and denser integration of small-scale producers within the regional and global market, particularly through the formation of export-oriented agro-industrial cluster zoning based on the methods and priorities of agribusiness, which over the past decade has stood as the fastest growing business sector.[44]

Food insecurity and technological fix

The development narrative of food security argues that the goal of ending hunger and malnutrition in Uganda can be achieved by strengthening governance, increasing investments and appropriately managing natural resources. A household is food secure if it can reliably gain access to food in sufficient quantity and quality for all its members to enjoy a healthy and active life.[45] The discourse of food security is intermeshed with a series of concerns linked to economies of scale, the nutritional content of staple crops and the purchasing power of households. In this context food insecurity seems to be caused by the lack of food availability at household level, rather than uneven distribution and access to land. This represents a significant shift in the traditional focus of food security from a question of political choice at the national level to one of purchasing power at the household level.

Despite the periodic outburst of food shortages across the country, the World Food Programme (WFP) argues that, overall, Uganda does not suffer from a lack of food supply.[46] There is in fact wider recognition in the country that the majority of food that circulates in the national market comes from small-scale agriculture. Thus the claim of food insecurity needs to be further investigated. Indeed, insufficient access to food is experienced by particular social categories in different places at different times. Food insecurity crosses both urban and rural spaces. Landless rural households or impoverished tenants, especially those in Uganda's southern region who have insufficient access to rural assets and resources, fail to produce their own subsistence requirements.[47] Moreover, farmers who are more integrated within commercial agriculture experience periods of food scarcity. In the North 20 years of protracted war, the forced reclusion of large numbers of Acholi peasants in Internally Displaced People (IDP) camps, and the influx of food aid from the USA has seriously affected both food production and distribution.[48]

The urban poor also suffer uncertainty in their access to food, as seen in the outburst of food riots and popular protests resulting from food inflation in 2011. Riots erupted in Kampala, Mbale and Gulu, where the convergence of poverty, unemployment and skyrocketing oil and food prices exasperated an already precarious social order. In May 2011 food inflation rose to 39.3%. The cost of a bunch of *matooke*,[49] for instance, rose from 9 Uganda shillings to around 30 shillings in September 2010. The overall food inflation in the financial year

2011–12 reached an astonishing 30.6%.[50] There is little doubt that small-scale rural producers suffered just as much. Yet the mainstream narrative of food (in) security misinterprets the causes of the food crisis, and portrays those economic and political actors that are arguably part of the problem as key to resolving the crisis.

The latest versions of this argument emphasise the need to stimulate more efficient mechanisms of food accessibility. These include developing and expanding internal and external markets; assisting the private sector to improve food processing, storage, marketing and distribution; promoting a well-coordinated system for collecting and disseminating information on food marketing and distribution; and securing food aid. Indeed, the neoliberal influence is clear, as the food problem is largely addressed in terms of non-perfectly functioning markets because of distortions as a result of state intervention or ineffective information systems. Yet this framework neglects important and historical questions of access to food, land and other natural resources.

Other preoccupations in the food debate include the pressing questions of population increase and the presumed low nutritional content of traditional crops.[51] And, as seen in other parts of the global South, corporate-led technological innovations and a focus on the market have been the well-expressed solutions to this Malthusian dilemma. In the summer of 2013 the Ugandan parliament debated the Bio-Safety and Bio-Technology Bill, which advocated the adoption of genetic manipulation of crop plants through molecular-assisted selection in lieu of conventional plant breeding. The intent was to produce higher yields, pest-resistance and nutritionally enhanced crops through the application of genetic engineering to agricultural production. Under the aegis of the Alliance for a Green Revolution in Africa (AGRA) and through the support of a cartel of transnational corporations (TNCs), the process is in its inception. Under the leadership of the National Agricultural Research Organization, agricultural scientists are implementing laboratory experiments on bio-fortification of staple food such as bananas and cassava; on the enhancement of vitamin A in finger millet and sorghum; and on testing genetically modified varieties of water-efficient maize and disease-resistant soya beans. The Water Efficient Maize for Africa project, led by the Kenya-based African Agricultural Technology Foundation, and funded by the Bill and Melissa Gates Foundation and USAID, has been implemented since 2008 in South Africa, Mozambique, Tanzania, Uganda and Kenya. The stated goal of the project is to increase yield stability, reduce hunger and improve the livelihoods of millions of Africans.

These events raised a series of ethical, cultural and ecological concerns. Agronomists at Makerere University, for example, argue that the introduction of genetically modified varieties of sorghum and finger millet in an open environment will most likely engender the pollination of all other indigenous varieties. The effect would presumably be a reduction in the bio-diverse patrimony and a standardisation of the different varieties of sorghum and millet grown for millennia by eastern African peasants. Below the surface of philanthropic interventions and policy narratives, the objective seems to be to incorporate African food production and consumption systems into global food chains by privatising the main source of production and increasing foreign control over them. In fact, the key goal of AGRA is to access African genetic wealth without sharing benefits with, nor recognising the role of, peasant communities who have been developing the cultivars for centuries.[52] This process, generally referred to as bio-piracy, is accompanied by a parallel process of patenting seeds, which results from adding one or more genes to the crop varieties that rural populations have developed for centuries. Sterilising and patenting seeds, whose global market is controlled by an oligopoly of a few TNCs, represents an additional step in the integration of seeds into the global value chains. Following the purchase of hybrid seeds an agreement is signed by the farmers stipulating them not to diffuse, multiply or exchange seeds. These agreements are aggressively promoted and monitored through a set of complex rules articulated by the World Trade Organization and the World Intellectual Property Organization. By such means institutions of global governance are pushing peasants, who have suffered from multiple acts of dispossession and bio-piracy, to remunerate the patent holders.

Civil society discourses and responses

Against this background civil society organisations (churches, environmental organisations, local and foreign NGOs, Community Based Organizations [CBOs]) have articulated competing, and sometimes overlapping, discourses based on the right to food. By mobilising themselves around issues such as human rights, food security, sustainability and community-based development, they have played a major role in policy achievements that favour the right to adequate food.[53] Codified in the United Nations Bill of Human Rights, the right to food is an individual and universal right of every human being. With the adoption of the Voluntary Guidelines on the Right to Food, Uganda stands as a pioneer in establishing a legal framework on the right to food, as well as an example of state–civil society cooperation within these guidelines. Based on these achievements, a network of NGOs and CSOs h has begun to apply these guidelines and monitor the Ugandan government's activities in fighting hunger and malnutrition.

Civil society organisations have also begun to mobilise actions and discourses around a set of collateral issues such as large-scale land acquisitions, GMOs, terms of exchange for agricultural commodities, conservation and management of natural resources, as well as intellectual property rights. With regard to land acquisitions, for example, the Uganda Land Alliance, an umbrella organisation grouping different social actors around issues of land rights, has pressed the government to respect the United Nations Voluntary Guidelines, and highlighted its illegal practices.

Following the successful example of Kenya, where social mobilisation against the adoption of pro-GMO legislation obtained a temporary interdiction, segments of civil society in Uganda raised debate and awareness among the population about the health, ecological and cultural implications of GMO adoption. In a similar vein civil society organisations successfully opposed the adoption by the Ugandan government of the Convention of International Union for the Protection of New Varieties of Plants (UPOV). UPOV, an intergovernmental organisation with headquarters in Geneva, was established in Paris in 1961 with the aim of codifying intellectual property rights for plant breeders and encouraging the development of new varieties of plants. UPOV grants the breeders not only the right to sell and produce propagating materials, but also the additional rights of reproduction (multiplication), conditioning for purpose of propagation, sale and stocking. By granting and protecting new plant variety rights (PVR), the organisation and its policies impose numerous conditions and limitations on farmers, while requiring the payment of royalties to the companies that retain the PVR.

The 1991 Rio Convention on Biodiversity introduced the principle of an equitable sharing of benefits between breeders and farmers. Because Uganda mostly depends on agriculture, civil society organisations argue that the UPOV Convention is not the right instrument for the legal protection of farmers since it is not compatible with the ways and means of the majority of the population. They suggested instead that the protection of local communities can best be achieved through the African Model Law (AML), which takes smallholders' interests into account by allowing them to continue their age-old right to exchange, use and sell seeds.

In the face of such challenges the Ugandan state has deployed an arsenal of coercion and control by targeting progressive NGOs that advocate land and food rights. Such state actions have included spying on and threatening engaged researchers, sympathetic lawyers and journalists. The result has been an increasing fragmentation of civil society interventions, the partial co-optation of organisations into political and legal frameworks, and subordination of these organisations to state censorship. Overall the capacity for civil society organisations to put pressure on state institutions depends on their ability to act in connection with popular constituencies. The success of the Save Mabira Forest Campaign in 2007 proves a case in point.[54] Yet, although the Mabira case has been widely referred to as an example of sustained protest by civil society groups, and as a deterrent to undemocratic decision making and profit-driven business initiatives around the commons, in-depth analysis highlights the complex webs of power that exist between the state, the business community and civil society, as well as the ambiguous role played by non-governmental organisations.[55]

Food sovereignty as peasant praxis?

Drawing from Borras and Franco's critical comments on the need to address the land dimension within the food sovereignty project,[56] this section explores the implications of land-based social relations on the instantiation of food sovereignty from the perspective of agrarian subjects in northern Uganda. Through an exploration of Acholi peasantries' historical and contemporary agricultural practices, and their struggles to maintain access to land and food production, I suggest that food sovereignty be read as a set of social, political and ecological practices aimed at ensuring local self-sufficiency and social reproduction. This section therefore historicises praxis to ground food sovereignty in the social, political and ecological practices of rural populations.

Historical studies based on linguistics, archaeology and paleo-ecology have brought to light the existence of complex and elaborated food-producing systems in the Great Lakes region.[57] Situated at the crossroad of various population movements, intermarriages and exchanges, the region became the place of an agricultural synthesis after 500 BCE. The successful integration of cereal agriculture and livestock, as well as the planting of root crops such as yams, oil palm, beans, gourds, cowpeas and leguminous plants, established the basis for consolidated patterns of settlements and soil fertility maintenance into the drier zones of the region.[58] The cumulative agricultural expertise inherited created the foundation for what Shoenbrun defined the 'roots of agricultural abundance'.[59]

Colonial enterprise aimed to push peasants to produce cotton in order to pay taxes and other kind of fees, and to be able to afford the imported European consumer goods that penetrated the colonial economy. Yet Acholiland was marginal in many ways to the early British colonial rule in the late 19th century. As a dry, geographically remote and sparsely populated area located far to the north of the country, it seemed of little interest to the commercial appetites of the colony.

A nation of agriculturalists and stockbreeders, Acholi neither show propensity to cash-crop production nor to labour migration into the southern region.[60] Land availability and the absence of individualised land tenure gave comparative stability to the 'traditional' socioeconomic order.[61] Rights in cultivation descended from father to son and were subordinated to the continued use of the soil and membership of the clan. Consistent with what Mafeje has argued, such forms of occupation did not imply permanent ownership but only granted usufruct rights.[62] Women accessed land through their husbands, although they were in charge of those plots dedicated to food crops near the house. Each wife or mature woman of the household maintained two fields of millet, one field of sesame, a small plot each of beans, pigeon peas, ground nuts and spinach.[63] Seed production and selection was performed by women who, after marriage, used to come with seeds brought from the father's household to begin cultivation. In a context of low population densities, low technological development, sparse settlements and distance from the main trading centres, the Acholi agro-pastoralist communities developed many forms of mutual help and collective labour that sharpened their political and socio-reproductive autonomy, and connected households and village. Mutual help, obligation to the group, and the centrality of the household and village were all emphasised. Material distinctions between people were of little importance. Relations of production were cooperative not antagonistic. Appropriation was of nature in the course of production, without social appropriation.[64] The use value of land and its importance were cemented in a common ethic and responsibility in the use of land.

Acholi peasants depended on family labour and had little incentive to produce cash crop surpluses. Planting millet, sesame, pigeon peas, sorghum and cassava in specific succession ensured an autonomous food supply. Pluri-activity, mixed cropping and shifting cultivation, combined with control over labour capacity and seed reproduction represented the pillars of the Acholi's food self-sufficiency. As reported in the 1967 Agricultural Census, sorghum and millet were regularly inter-cropped, and often mixed with groundnuts, field and pigeon peas, and sim sim.[65] Poly-cultures based on a combination of sorghum, peanut and millet yielded consistently more that corresponding monocultures in conditions of stressful access to water.[66]

These dynamics and forms of agency acquire further relevance in the contemporary context of mounting competition over land resources, neoliberal restructuring of agriculture, prolonged war, the internment of 1.8 million people in IDP camps, cattle dispossession by Ugandan military elites, land grabs, and the massive influx of US food aid. Yet, in the face of continuous pressures, the northern regions are still by far the largest producers (in terms of hectares and production) of food crops.[67] These include sorghum, cow and pigeon peas, groundnuts and sim sim, while the regions are the second largest producer of finger millet, beans and cassava. In particular, the Amuru district is the largest producing district in the northern region for beans (74,671 tons) and groundnuts (14,375 tons).[68] Obviously these statistics do not provide information on the (re) distribution of food crop production among rural households, nor the disposition of family labour product. With regard to the latter, it is critical to highlight that only 19% of finger millet produced is sold, as the majority of the harvest is consumed (37.7%), stored (33.5%) and used for 'ceremonial' or reciprocity purposes (9.8%). Similarly, only 14.3% of sorghum is sold, while 46.9% is consumed, 30.1% is stored, and 8.6% is used for redistributive socio-cultural activities.[69]

Fieldwork in Amuru confirms these data. There were minimal levels of commercialisation of agricultural outputs, although some variation exists between crops. The district has also been characterised by very low levels of access to formal credit (1.5%) or to government agricultural modernisation programmes like National Agricultural Advisory Services (NAADS) (1.2%).[70] In the absence of support programmes from above, peasants seem to rely on more horizontal networks (farmer-to-farmer) of knowledge transmission in regard to agricultural techniques, crop varieties, conservation practices and climatic conditions. In terms of land access northern Uganda still enjoys the highest size of average holding per capita, 1.6 hectares, compared to the national average, 1.1 hectares.[71] In Amuru 98.4% of people access land through customary land-tenure systems, while freehold registration accounts for only 0.5% of parcels.[72] Notwithstanding the absence of written documentation, only 0.3% of respondents in the area perceived this form of tenure as insecure.[73] In fact, the low levels of commercialisation, relative availability of land through customary land tenure systems, and the presence of vibrant local peasant markets in the region have thwarted the entrenchment of commodity relations. As Harriet Friedmann notes, 'If household reproduction rests on reciprocal ties, then reproduction resists commoditization. If access to land, labour and credit and product markets, is mediated through direct non monetary ties to other households or classes...then commodity relations are limited in their ability to penetrate the cycle of reproduction.'[74]

This is an incisive reminder that, although capitalism is the dominant mode of production, where peasant households' reproduction rests on a set of communal and class relations, there are objective limitations to the operation of the law of value. This does not mean that Acholi is an egalitarian society, nor that peasant households are insulated from broader market imperatives. Rather, it indicates that, while social stratification is rapidly advancing, and class formation is crystallising in the form of a tripartite division into peasant agriculture, entrepreneurial agriculture and agribusiness,[75] peasants' forms of social organisation are actively shaping the trajectories of agrarian change.

Yet this social group faces enormous challenges emanating from other social groups which represent the interests of traditional-turned-entrepreneurial, agrarian classes – the village bourgeoisie.[76] They are receptors and implementers of the precepts and impulses deployed by agribusiness: mechanisation, production for export markets, chemicalisation of agriculture and workers' over-exploitation. In doing so, they are the promoters of a view that increasingly sees land and agriculture as sources of profit, and are often invested in other remunerative activities in the countryside, such as merchant activities, transport services and as landlords.

In capitalist agriculture all resources are viewed as commodities, including labour.[77] In Amuru labour commoditisation is impeded by reciprocity between family, households and community. Similar exchange within the community occurs with other resources such as seeds. It is therefore crucial to assume the existence of different degrees and forms of commoditisation, which radically alter the spectrum of possibilities and opportunities for rural households. Even when peasants respond to market opportunities, they act differently from market driven actors in respect to profit maximisation and accumulation. For peasants subsistence and survival drive agricultural production, not the 'agricultural entrepreneurial calculus'.[78] By determining the crop varieties to plant, the seeds to use, the quantities to sell, and by constructing social relations that counter market consumption, households can play a significant role in shaping the impact of the market within their local area.[79] For instance, rural households have been successful in protecting their capacity to reproduce seeds in the face of corporate appropriation of genetic resources and seed patenting, and the attempts by NGOs to promote hybrid seed varieties among farmers in the area. Seed selection and reproduction is reserved as the meticulous role of women, who have produced a veritable set of local seed banks. Seeds also represent an important means of social exchange. The elaborated nature of local food systems and their embeddedness in socio-ecological relations are of paramount significance here to explain the resilience of peasant farming in a context of mounting pressure for commoditisation. These insights into the social reproduction strategies of Acholi peasants illuminate the character of social practices that underpin food sovereignty.

The recent and ongoing struggle against the attempt by the Ugandan state to grab 40,000 ha of land in the Amuru district and allocate it to the Madhvani Group for large-scale sugar plantation – a resistance led by women – highlights the great degree of awareness among rural populations regarding the importance of maintaining access to reproductive resources. In other cases, where dispossession occurred in the district to make space for conservation areas and game reserves, peasants re-enacted strategies of mobile farming at the margins of enclosed lands. The resurfacing of hidden forms of contestation reinvigorate the significance of the land and food question, especially in a post-conflict context where precarious and insecure existences are the norm, and peasants continue to value their access to common land deeply as the main avenue to social reproduction.

Concluding remarks and future research trajectories

Paraphrasing the first lines of Marx's *The Eighteenth Brumaire of Louis Bonaparte*, McMichael argued that 'peasants make their own history, but not always as they please'.[80] This article has argued that many of the organising principles at the core of the food sovereignty paradigm are inscribed in the socio-cultural and ecological practices of peasant populations in northern Uganda. Yet these practices are taking place in an increasingly adverse national and international environment, and under circumstances transmitted from the past, which challenge their implementation enormously and jeopardise the future of food security/sovereignty prospects for peasant agriculture. The article has explored this tension by suggesting the need to understand food sovereignty from the everyday peasant practices which embody an incredibly valuable, historically constructed and territorially grounded knowledge. This might help to move the debate beyond the epiphenomenal character of its contemporary formulations and incorporate its historical and localised meaning. Such insights need to be put into dialogue with global formulations elaborated by transnational agrarian movements (TAMs), which have been instrumental in conceiving and promoting the notion of food sovereignty.[81] TAMs have not occupied a central political stage in Africa as they have in Latin America and Southeast Asia. African social movements did gain some momentum, especially as a result of the African social forums in Nyéléni (2007), Maputo (2008), and again in the anti-land grabbing conference held in Nyéléni (2011). Peasant organisations, however, have remained extremely fragmented and disconnected from the wider political arena. Major limitations include the difficulty of intercepting localised and everyday forms of rural struggles and connecting them to larger more comprehensive demands.

Local instances, practices and struggles represent the substratum upon which global interconnections are drawn. It is along the local–global axes that the dynamics of capital accumulation and resistance take place. It is along this axis that future research needs to concentrate. The nucleus of the future research trajectory should include in-depth explorations of the changing class structures and agencies of the Ugandan countryside in order to grapple with the character of everyday politics. Particular emphasis should be given to the dynamics of production, consumption, distribution and social reproduction at the household level. In such a phase of accelerated capitalist penetration and transformation, many aspects of the food sovereignty question remain unknown. It is essential to devote particular attention to changing dynamics of land use, seed production, consumption patterns, the sexual division of labour, use of fertilisers and pesticides, and water access. This requires a close investigation of the singular conditions of constraint and opportunity that different categories of small-scale farmers confront in particular space–time settings.

Acknowledgements

I want to thank Arthur Owor Spender for fieldwork assistance and the guest editors of this special issue, who provided insight and expertise that greatly improved the manuscript. Last but not least I need to thank the residents of the villages in Amuru District who shared their time and knowledge, and without whom this work would not have been possible. To them this work is dedicated.

FOOD SOVEREIGNTY

Notes

1. Bush, "Food Riots."
2. United Nations, *World Economic Situation*, 7.
3. FAO, *Hunger on the Rise*.
4. McMichael, "Food Sovereignty in Movement"; and Bello, *The Food Wars*.
5. Weis, "The Accelerating Biophysical Contradictions"; and Weis, "The Meat of the Global Food Crisis."
6. Araghi, "The Invisible Hand," 133.
7. Desmarais, *La Vía Campesina*.
8. Witman et al., "The Origins and Potential," 2.
9. McMichael, "Food Sovereignty, Social Reproduction."
10. Kloppenburg, "Impeding Dispossession, Enabling Repossession," 372.
11. Altieri and Nicholls, "Scaling up Agroecological Approaches," 121.
12. Van der Ploeg, "Peasant-driven Agricultural Growth," 1000.
13. Agarwal, "Food Sovereignty, Food Security."
14. Bernstein, "Food Sovereignty via the 'Peasant Way'."
15. Edelman, "Food Sovereignty."
16. Boyer, "Food Security, Food Sovereignty," 333.
17. Bryceson et al., *Disappearing Peasantries?*
18. Bernstein, *Class Dynamics of Agrarian Change*.
19. Agarwal, "Food Sovereignty, Food Security," 1252.
20. Ratjen et al., "Food Sovereignty and Right to Food."
21. UBOS, *2012 Statistical Abstract*, 41.
22. Ibid., 224.
23. Ibid., v.
24. De Soto, *The Mystery of Capital*.
25. Mailo land is the outcome of the 1900 Buganda Agreement between the British Crown and the Baganda Oligarchy, which assigned exclusive control of 9000 square miles to Baganda traditional chiefs.
26. Batungi, *Land Reform in Uganda*.
27. Mamdani, "The Contemporary Ugandan Discourse."
28. McAuslan, *Land Law Reforms in East Africa*.
29. Manji, *The Politics of Land Reform in Africa*.
30. Lahiff et al., "Market-led Agrarian Reform."
31. World Bank, *Uganda Economic Update Special Focus*, 57.
32. See Mitchell, "The Properties of Markets."
33. Mamdani, *Politics and Class Formation*; and Tosh, "Lango Agriculture."
34. Mamdani, "Peasants and Democracy in Africa."
35. Bategeka et al., *Institutional Constraints to Agricultural Development*, 2.
36. Okello, *Opposition Response to the Government Budget*, 13.
37. FAO, *WTO Agreement on Agriculture*.
38. IFPRI. *The Supply of Inorganic Fertilizers*, 4.
39. World Bank, *Uganda Economic Update*, xiii–xvi.
40. World Bank, *Uganda*.
41. World Bank, *Uganda Economic Update*, 31.
42. EPRC, *Improving the Use of Agricultural Technologies*, 1.
43. Museveni, "State of the Nation," 3.
44. World Bank, *Uganda Economic Update*, 22, 58.
45. Republic of Uganda, *The National Food and Nutrition Strategy*, iv.
46. World Food Programme, *Uganda Overview*.
47. Lwanga-Lunyiigo, *The Struggle for Land*.

48. Branch, *Displacing Human Rights*.
49. The most common staple food in Uganda, made out of bananas.
50. World Bank, *Uganda Economic Update*, 66.
51. Republic of Uganda, *The National Food and Nutrition Strategy*.
52. Thompson, "Alliance for a Green Revolution," 345
53. Ratjen et al., "Food Sovereignty and Right to Food," 27.
54. Child, "Civil Society in Uganda."
55. Hönig, "Civil Society and Land Use."
56. Borras and Franco, "Food Sovereignty & Redistributive Land Policies," 107.
57. Schoenbrun, *A Green Place, A Good Place*; and Schoenbrun, "We are what we Eat."
58. Schoenbrun, *A Green Place, A Good Place*, 20.
59. Schoenbrun, "We are what we Eat," 65.
60. Atkinson, *The Roots of Ethnicity*, 5.
61. Girling, *The Acholi of Uganda*, 183.
62. Mafeje, *The Agrarian Question*.
63. Girling, *The Acholi of Uganda*, 191.
64. Mamdani, *Politics and Class Formation in Uganda*, 21.
65. Uganda Government, *Report on Uganda Census of Agriculture*.
66. Natarajan and Willey, quoted in Altieri, "Scaling up Agroecological Approaches," 125.
67. UBOS, *2012 Statistical Abstract*, 43.
68. Ibid., 42.
69. Ibid., 44.
70. UBOS, *Uganda Census of Agriculture 2008/2009, Vol. III*, 304, 310, 323.
71. UBOS, Ibid., *Vol. IV*, 14.
72. Ravnborg et al., *Land Tenure under Transition*, 17.
73. Ibid., 22, 39.
74. Friedmann, "Household Production and the National Economy," 163.
75. See Van der Ploeg, *The New Peasantries*, 1–6.
76. See Mamdani, *Politics and Class Formation in Uganda*.
77. Van der Ploeg, "Peasant-driven Agricultural Growth," 1004; and Wood, "Peasants and the Market Imperative."
78. Friedmann, "Household Production and the National Economy."
79. Bush, *Poverty and Neoliberalism*, 194.
80. McMichael, "Peasants make their own History."
81. Borras et al., *Transnational Agrarian Movements*.

Bibliography

Agarwal, Bina. "Food Sovereignty, Food Security and Democratic Choice: Critical Contradictions, Difficult Conciliations." *Journal of Peasant Studies* 41, no. 6 (2014): 1247–1268. doi:10.1080/03066150.2013.876996.

Altieri, Miguel, and Clara Nicholls. "Scaling up Agroecological Approaches for Food Sovereignty in Latin America." In *Food Sovereignty: Reconnecting Food, Nature and Community*, edited by Hannah Kay Wittman, Annette Aurelie Desmarais and Nettie Wiebe, 120–133. Oakland, CA: Food First, 2010.

Araghi, Farshad. "The Invisible Hand and the Visible Foot." In *Peasants and Globalization: Political Economy, Rural Transformation and the Agrarian Question*, edited by Haroon Akram-Lodhi and Cristóbal Kay, 111–147. London: Routledge, 2009.

Atkinson, Ronald R. *The Roots of Ethnicity: Origins of the Acholi of Uganda before 1800*. Kampala: Fountain Publishers, 2010.

Bategeka, Lawrence, Julius Kiiza, and Ibrahim Kasirye. *Institutional Constraints to Agricultural Development in Uganda*. Kampala: Economic Policy Research Centre, 2013.

Batungi, Nasani. *Land Reform in Uganda: Towards a Harmonised Tenure System*. Kampala: Fountain Publishers, 2008.

Bello, Walden. *The Food Wars*. London: Verso, 2009.

Bernstein, Henry. "Food Sovereignty via the 'Peasant Way': A Skeptical View." *Journal of Peasant Studies* 41, no. 6 (2014): 1031–1063. doi:10.1080/03066150.2013.852082.

Bernstein, Henry. *Class Dynamics of Agrarian Change*. Halifax: Fernwood Publishing, 2010.

Borras, Saturnino M. Jr., and Jennifer C. Franco. "Food Sovereignty & Redistributive Land Policies: Exploring Linkages, Identifying Challenges." In *Food Sovereignty: Reconnecting Food, Nature and Community*, edited by Hannah Kay Wittman, Annette Aurelie Desmarais and Nettie Wiebe, 106–119. Oakland, CA: Food First, 2010.

Borras, Saturnino M. Jr, Marc Edelman, and Cristóbal Kay. eds. *Transnational Agrarian Movements Confronting Globalization*. Chichester: Wiley-Blackwell, 2008.

Boyer, Jefferson. "Food Security, Food Sovereignty, and Local Challenges for Transnational Agrarian Movements: The Honduras Case." *Journal of Peasant Studies* 37, no. 2 (2010): 319–351. doi:10.1080/03066151003594997.

Branch, Adam. *Displacing Human Rights: War and Intervention in Northern Uganda*. Oxford: Oxford University Press, 2011.

Bryceson, Deborah Fahy, Cristóbal Kay, and Jos Mooji. *Disappearing Peasantries? Rural Labour in Africa, Asia and Latin America*. London: Intermediate Technology, 2000.

Bush, Ray. "Food Riots: Poverty, Power and Protest." Journal *of Agrarian Change* 10, no. 1 (2010): 119–129. doi: 10.1111/j.1471-0366.2009.00253.

Bush, Ray. *Poverty and Neoliberalism: Persistence and Reproduction in the Global South*. London: Pluto Press, 2007.

Child, Keith. "Civil Society in Uganda: The Struggle to save the Mabira Forest Reserve." *Journal of Eastern African Studies* 3, no. 2 (2009): 240–258.

Desmarais, Annette Aurelie. *La Vía Campesina: Globalization and the Power of Peasants*. London: Pluto Press, 2007.

De Soto, Hernando. *The Mystery of Capital: Why Capitalism triumphs in the West and fails Everywhere Else*. London: Black Swan, 2000.

Edelman, Mark. "Food Sovereignty: Forgotten Genealogies and Future Regulatory Challenges." *Journal of Peasant Studies* 41, no. 6 (2014): 959–978. doi:10.1080/03066150.2013.876998.

EPRC. *Improving the Use of Agricultural Technologies in Uganda*. Policy Brief no. 27. Kampala: Economic Policy Research Centre, 2013.

FAO. *Hunger on the Rise*. Briefing Paper. New York: United Nations, 2008.

FAO. *WTO Agreement on Agriculture: The Implementation Experience – Developing Country Case Studies*. Rome: FAO, 2003.

Friedmann, Harriet. "Household Production and the National Economy: Concepts for the Analysis of Agrarian Formations." *Journal of Peasant Studies* 7, no. 2 (1980): 158–184.

Girling, F. *The Acholi of Uganda*. London: HMSO, 1960.

Hönig, Patrick. "Civil Society and Land Use Policy in Uganda: The Mabira Forest Case." *Africa Spectrum* 49, no. 2 (2014): 53–77.

IFPRI. *The Supply of Inorganic Fertilizers to Smallholder Farmers in Uganda*. Policy Note 16. Washington, DC: International Food Policy Research Institute, 2013.

Kloppenburg, Jack. "Impeding Dispossession, Enabling Repossession: Biological Open Source and the Recovery of Seed Sovereignty." *Journal of Agrarian Change* 10, no. 5 (2010): 367–388.

Lahiff, Edward. "Saturnino M. Borras Jr., and Cristóbal Kay. "Market-led Agrarian Reform: Policies, Performance and Prospects." *Third World Quarterly* 28, no. 8 (2007): 1417–1436.

Lwanga-Lunyiigo, Samwiri. *The Struggle for Land in Buganda 1888–2005*. Kampala: Wavah Books, 2007.

Mafeje, Archie. *The Agrarian Question, Access to Land, and Peasant Responses in Sub-Saharan Africa*. Geneva: United Nations Research Institute for Social Development, 2003.

Mamdani, Mahmood. "The Contemporary Ugandan Discourse on Customary Tenure: Some Historical and Theoretical Considerations." Paper presented at the Workshop 'The Land Question: Socialism, Capitalism and the Market', Makerere Institute of Social Research, Kampala, August 9–10, 2012.

Mamdani, Mahmood. "Peasants and Democracy in Africa." *New Left Review* I, no. 156 (1986): 37–49.

Mamdani, Mahmood. *Politics and Class Formation in Uganda*. New York: Monthly Review Press, 1976.

Manji, Ambreena. *The Politics of Land Reform in Africa: From Communal Tenure to Free Markets*. London: Zed Books, 2006.

McAuslan, Patrick. *Land Law Reforms in East Africa: Traditional or Transformative?* London: Routledge, 2013.

McMichael, Philip. "Food Sovereignty in Movement: Addressing the Triple Crisis." In *Food Sovereignty: Reconnecting Food, Nature and Community*, edited by Hannah Kay Wittman, Annette Aurelie Desmarais and Nettie Wiebe, 168–185. Oakland, CA: Food First, 2010.

McMichael, Philip. "Food Sovereignty, Social Reproduction and the Agrarian Question." In *Peasants and Globalization: Political Economy, Rural Transformation and the Agrarian Question*, edited by Haroon A. Akram Lodhi and Cristóbal Kay, 288–312. London: Routledge, 2009.

McMichael, Philip. "Peasants make their own History, but not just as they Please..." *Journal of Agrarian Change* 8, nos. 2–3 (2008): 205–228.

Mitchell, Timothy. "The Properties of Markets." In *Do Economists make Markets? On the Performativity of Economics*, edited by Donald Mackenzie, Fabian Muniesa and Lucia Siu, 244–275. Princeton, NJ: Princeton University Press, 2007.

Museveni, Yoweri. "State of the Nation." Address at the opening of the 2nd session of the ninth parliament. Kampala, 2012.

Okello, Okoman A.C. *Opposition Response to the Government Budget Statement FY 2009/10*. Kampala, 2009.

Ratjen, Sandra, Irene Wasike Muwanguzi, Peter Rukundo, and James Kintu. "Food Sovereignty and Right to Food: The Case of Uganda." *Rural21* 42, no. 3 (2008): 25–27.

Ravnborg, Helle Munk, Bernard Bashaasha, Rasmus Hundsbæk Pedersen, Rachel Spichiger, and Alice Turinawe. *Land Tenure under Transition: Tenure Security, Land Institutions and Economic Activity in Uganda*. Copenhagen: Danish Institute for International Studies, 2013.

Republic of Uganda. *The National Food and Nutrition Strategy*. Kampala: Ministry of Agriculture, Animal Industry and Fisheries, 2005.

Schoenbrun, David L. *A Green Place, A Good Place: Agrarian Change, Gender and Social Identity in the Great Lakes Region to the 15th Century*. Kampala: Fountain Publishers, 1998.

Schoenbrun, David L. "We are what we Eat: Ancient Agriculture between the Great Lakes." *Journal of African History* 34, no. 1 (1993): 1–31.

Thompson, Carol B. "Alliance for a Green Revolution in Africa (AGRA) – Advancing the Theft of African Genetic Wealth." *Review of African Political Economy* 39, no. 132 (2012): 345–350.

Tosh, John. "Lango Agriculture during the Early Colonial Period: Land and Labour in a Cash-crop Economy." *Journal of African History* 19, no. 3 (1978): 415–439.

United Nations. *World Economic Situation and Prospects*. New York, NY: United Nations, 2009.

UBOS. *2012 Statistical Abstract*. Kampala: Republic of Uganda, 2012.

UBOS. *Uganda Census of Agriculture 2008/2009. Vol. III: Agricultural Household and Holding Characteristics Report*. Kampala: Republic of Uganda, 2010.

UBOS. *Uganda Census of Agriculture 2008/2009. Vol. IV: Crop Area and Production Report*. Kampala: Republic of Uganda, 2010.

Uganda Government. *Report on Uganda Census of Agriculture*. Entebbe, 1967.

Van der Ploeg, Jan Douwe. "Peasant-driven Agricultural Growth and Food Sovereignty." *Journal of Peasant Studies* 41, no. 6 (2014): 999–1030. doi:10.1080/03066150.2013.876997.

Van der Ploeg, Jan Douwe. *The New Peasantries: Struggles over Autonomy and Sustainability in the Era of Empire and Globalization*. London: Earthscan, 2008.

Weis, Tony. "The Meat of the Global Food Crisis." *Journal of Peasant Studies* 40, no. 1 (2013): 65–85.

Weis, Tony. "The Accelerating Biophysical Contradictions of Industrial Capitalist Agriculture." *Journal of Agrarian Change* 10, no. 3 (2010): 315–341.

Wittman, Hannah Kay, Annette Aurelie Desmarais, and Nettie Wiebe. "The Origins and Potential of Food Sovereignty." In *Food Sovereignty: Reconnecting Food, Nature and Community*, edited by Hannah Kay Wittman, Annette Aurelie Desmarais and Nettie Wiebe, 1–14. Oakland, CA: Food First, 2010.

Wood, Ellen Meiksins. "Peasants and the Market Imperative: The Origins of Capitalism." In *Peasants and Globalization: Political Economy, Rural Transformation and the Agrarian Question*, edited by Haroon A. Akram Lodhi and Cristóbal Kay, 37–56. London: Routledge, 2009.

World Bank. *Uganda Economic Update Special Focus: Jobs – Key to Prosperity*. Washington, DC: World Bank, 2013.

World Bank. *Uganda: Promoting Inclusive Growth*. Washington, DC: World Bank, 2012.

World Food Programme. *Uganda Overview*. 2011. www.wfp.org/countries/Uganda/Overview.

Challenges for food sovereignty policy making: the case of Nicaragua's Law 693

Wendy Godek

Division of Global Affairs, Rutgers University, Newark, USA

Food sovereignty policy initiatives face significant challenges in their quest to be approved. This article examines the case of Nicaragua's Law 693, the Law of Food and Nutritional Sovereignty and Security, which was passed in 2009. Drawing on empirical research, the article details the initial stages of the policy-making process – from the origins and development of the proposal for a food sovereignty law to its introduction and initial deliberation by the National Assembly to the breakdown in the approval process because of conflict over the law's content. Using theoretical insights from the food sovereignty and food security policy literature, Law 693 is examined, noting key limitations food sovereignty faced during the policy-making process. The study finds that the strength and force of national food sovereignty discourses, the ability of food sovereignty advocates to convince others of the legitimacy and viability of the food sovereignty approach, and the willingness of the state to create the necessary conditions to foster food sovereignty are all important factors when evaluating the potential for food sovereignty to be successfully adopted into public policies.

Introduction

Food sovereignty can be briefly defined as the 'right of nations and peoples to control their own food systems, including their own markets, production modes, food cultures and environments'.[1] The origin of the term can be traced back to a Mexican government food programme in the early 1980s and was later adopted – albeit to a limited extent – by Central American peasant activists in the mid- to late 1980s and early 1990s.[2] Some of these groups went on to become founding members of La Vía Campesina (LVC),[3] which presented the concept of food sovereignty at the 1996 World Food Summit (WFS) as a response to the deepening crisis in rural communities spurred by the expansion of the neoliberal development project and the globalisation of both the dominant productionist approach to agriculture and food (agrifood) systems and the market-based approach to food security.[4] LVC viewed these market-based solutions to food insecurity as insufficient for addressing the underlying structural causes of poverty and hunger.[5] It instead forwarded the alternative paradigm of food sovereignty, arguing it to be a 'precondition for genuine food security'.[6]

A fundamental goal of the food sovereignty movement has been the institutionalisation of food sovereignty through broad policy change grounded in participatory and democratic policy-making processes. In the late 1990s there was turn towards pursuing national food sovereignty policies. Such proposals were met with success in Venezuela, Mali, Senegal, Nepal, Bolivia, Ecuador, Nicaragua and, most recently, the Dominican Republic.[7] The adoption of food sovereignty policies, however, has not come without struggle on the part of the social movements and civil society organisations (CSOs) that support them.[8] This is unsurprising given the nature of food sovereignty, which both critiques the dominant approach to agrifood systems and at the same time forwards a multidimensional alternative that emphasises, among other principles, the right to food for all, food democracy, agro-ecological production, and reform of the neoliberal trade regime.[9]

This article explores challenges faced by food sovereignty policy proposals by looking at the case of Nicaragua's Law 693, the Law of Food and Nutritional Sovereignty and Security. Drawing on empirical research conducted in Nicaragua, it examines an important segment of the process by which Law 693 was made: the development of the draft law, which was introduced to the National Assembly in October 2006, to the breakdown in the approval process in June 2007 as a result of conflict over the content of the bill, namely food sovereignty-related provisions. The law was eventually passed in 2009 following a lengthy process of intense multiple stakeholder negotiations, although food sovereignty was significantly weakened in the revised final version. While this article does not analyse the entire policy-making process of Law 693, it illustrates how, despite political opportunities and other favourable factors, food sovereignty faced important constraints from the outset. While these challenges are not unique to the Nicaraguan case, this example is valuable for examining how such challenges manifest themselves in different contexts.

To assist in unpacking the story of Law 693, the section that follows briefly discusses several theoretical challenges for food sovereignty policies. The article then turns to the case study, beginning with a short history of the Nicaraguan food sovereignty movement and food security policy in Nicaragua, and then continues with a discussion of the initial stages of the policy-making process of Law 693. It closes with a discussion of the case study in terms of the earlier challenges, thus offering a brief reflection on important aspects of food sovereignty policy making in theory and practice.

Challenges to food sovereignty policy making

The promotion of policies has been one way that the food sovereignty movement has broadly sought to institutionalise food sovereignty as a means of achieving lasting food security and realising the right to food. However, because of controversy over the types of changes envisioned under the food sovereignty framework, such policy proposals face different types of challenge as they move through the policy-making process. In the following, three significant factors that influence the successful and meaningful institutionalisation of food sovereignty into policy are discussed; they include competition from other approaches to food security, the conceptual ambiguity of food sovereignty, and the paradox of the state. Together these challenges provide a lens through which to analyse the initial stages of Law 693's policy-making process.

Competing approaches to food security

Food sovereignty is one approach to achieving the goal of food security and thus faces competition from other views on how to accomplish this goal.[10] Calling food sovereignty a 'counterframe' to trade-based food security, Fairbairn invoked this idea of competing approaches to food security. Indeed, frame theory is useful for conceptualising such conflict. As argued by Moony and Hunt, food security functions as 'a particularly potent form of master frame',[11] which represents what Gamson called a 'consensus frame',[12] defined as 'a concept or term that finds broad resonance and consent, but which is used to make diverging, and sometimes conflicting claims'.[13] Comprising the food security consensus frame are different collective action frames that often clash in their visions of how to achieve the broader goal of food security because of the distinct interests of relevant actors. A noted example here is the trade-based approach versus the food sovereignty approach.[14] Applied to policy, the positions of food security stakeholders reflect the respective collective action frames to which they subscribe and are particular to the specific policy context.[15] The policy-making process is an important site in which these conflicting food security policy frames compete against each other as policy stakeholders wield their power to influence a final outcome favourable to their interests. It is here that the food sovereignty frame must have force in terms of providing a compelling approach that persuades others of its legitimacy *vis-à-vis* its competitors.

The conceptual ambiguity of food sovereignty

Within this context of competing approaches to food security food sovereignty suffers from another weakness: its lack of conceptual clarity, or what critical discourse analysts might call a lack of discursive coherence.[16] The premise here is that the absence of a common articulation and collective understanding of what food sovereignty means points to conceptual ambiguity, which can leave the concept open to misinterpretation and thus affect the way in or extent to which the concept and its framework are appropriated by policy – or otherwise. Some of the major features of the challenge, as Patel has pointed out, include contradictions inherent in the discourse and the multiplicity of food sovereignty definitions.[17] Moreover, as Wittman and Desmarais have suggested, the existence of multiple interpretations of food sovereignty within the movement and among different groups can have important implications for the successful formation of local or national food sovereignty movements and food sovereignty policy making in general.[18] A final and important source of confusion is food sovereignty's reframing of the political concept of sovereignty – usually applied to nation-states – to refer to the self-determination and autonomy of individuals and groups at multiple scales (local, territorial, regional, national, etc).[19] Boyer argued that 'the idea of autonomy evoked by the term sovereignty ... can become somewhat confusing to the many who equate sovereignty with states and with the rights of particular peoples or aspects of their daily lives'.[20] The results of his research suggest that the semantics of the word 'sovereignty' can be unclear because of its 'successive "stacking" of multiple meanings'; the 'complexity' of the food sovereignty concept may be one of the factors that limits its appropriation.[21]

The paradox of the state

A final challenge is the contradiction posed by the state in food sovereignty policy making. As Clark has observed, the 'state is both part of the problem and the solution for achieving greater [food sovereignty]'; for example, on the one hand, the state can hinder food sovereignty policies by upholding policies antithetical to the framework, while on the other hand, policy space needs to be made by the state in order to achieve food sovereignty reforms.[22] Along these lines McKay et al have depicted the state as a decisive factor in creating the structural changes envisioned by the food sovereignty framework from the moment of policy formulation forward, particularly with respect to honouring the core principle of food sovereignty – the people's right to define their own food and agriculture policies.[23] Schiavoni has argued that 'the adoption of food sovereignty into state policy ... calls for a redefining of the terms of engagement between state and society'.[24] This is particularly important because the bottom-up, participatory policy-making style demanded by food sovereignty is quite nascent in practice.[25]

Food sovereignty in Nicaragua and the case of Law 693

Having outlined a brief theoretical framework, the article now turns to the case study of Law 693. It begins with a short overview of the roots of the food sovereignty movement in Nicaragua and early food security policy efforts to historically situate the proposal for food sovereignty legislation. It then continues by recounting the initial steps of the policy-making process towards approving food security legislation that included the concept and principles of food sovereignty.

Origins of the Nicaraguan food sovereignty movement

While an in-depth discussion of the roots of food sovereignty in Nicaragua is beyond the scope of this article, several developments occurred during the Revolution, led by the National Sandinista Liberation Front (FSLN), that laid the foundation for the emergence of the Nicaraguan food sovereignty movement and influenced its contours. First, policies, ideologies and rhetoric emerged with the Revolution that emphasised sovereignty and autonomy,[26] pluralist democracy with citizen participation at multiple scales of governance, food security and self-sufficiency through domestic production. In addition, there was a focus on creating greater access for all to productive resources, especially formerly marginalised peasants and small and medium producers, in order to stimulate rural and peasant production. These were particularly visible in the successive stages of Agrarian Reform from 1979 to 1986 and in the 1982 National Food Policy (PAN), which was the first food security policy of its kind in Central America.[27] Second, the Revolutionary period was characterised by transnational exchanges as foreigners came to Nicaragua to witness the Revolutionary experiment and express their solidarity with the Revolution and Nicaraguans. This resulted in dialogues between Nicaraguan peasant and farmer organisations and other regional and international peasant and farmer organisations, as well as international NGOs and cooperation organisations, ultimately fostering solidarity between these groups and the sharing of ideas and information. Third, the exchange of ideas and experiences between Nicaraguans and foreigners resulted in the emergence of new national movements in the 1980s – such as the Farmer-to-Farmer Programme (PCAC) and the Nicaraguan Environmental Movement (MAN) – which promoted traditional production systems and ecologically mindful forms of production.[28] These were a response to economically and environmentally unsustainable industrial models of agriculture advocated under the agrarian reform. As a result of these developments, certain conditions were created that allowed for further national and transnational mobilisation in the early 1990s. Third, the exchange of ideas and experiences between Nicaraguans and foreigners resulted in the emergence of new national movements in the 1980s in response to economic and environmentally unsustainable industrial model of agriculture advocated under the agrarian reform – such as the Farmer-to-Farmer Program (PCAC) and the Nicaraguan Environmental Movement (MAN) – that promoted traditional production systems and ecologically-mindful forms of production.[29] As a result of these developments, certain conditions were created that allowed for further national and transnational mobilization in the early 1990s.

The Rural Workers Union (ATC) and the National Farmers and Ranchers Union (UNAG), Nicaraguan peasant and farmer organisations with their roots in the Revolution, were among the founding members of LVC and participated in the development of the food sovereignty concept within this transnational space. Parallel to the development of the transnational food sovereignty movement, a national movement for food sovereignty spearheaded by the ATC and UNAG began to emerge in Nicaragua in the 1990s in response to the effects of far-reaching political and economic changes with profound implications for rural communities. The Sandinista loss in the 1990 general election brought a neoliberal regime to power that immediately introduced structural adjustment reforms. As a result, hallmark Revolutionary reforms, including agrarian reform and PAN, were dismantled; export production was prioritised over production for domestic markets; migration both to urban centres and abroad ensued as a result of unemployment; and the opening of the market led to an increase in transnational investment. In consequence peasant production suffered and access to food became a problem as the logic changed from producing one's own to relying on imports. Faced with the threats generated by the open market, the concepts of sovereignty and autonomy re-emerged as significant.[30]

In light of these challenges, early efforts to promote food sovereignty focused on addressing the national rural crisis by putting food sovereignty principles into practice at the local level. These initiatives were influenced by Revolutionary experience and ideology, particularly the focus on national food autonomy, as well as by the growing focus on sustainable agricultural production, which was attributable to the early efforts of PCAC and MAN in the late 1980s, followed by the establishment of the Group for the Promotion of Ecological Agriculture (GPAE) in 1994 by civil society and farmer organisations and by the ongoing efforts of PCAC. During the 1990s different projects and programmes encouraging local production for consumption, such as the Food Production Voucher (BPA) programme, which provided families with productive inputs at the municipal level, were largely undertaken by the ATC, PCAC, the National Union of Associated Agricultural Producers (UNAPA), sympathetic nongovernmental organisations and municipal governments and, upon its creation in the late 1990s, the Agriculture and Forestry Roundtable (MAF).[31] The objectives of these initiatives were to strengthen family-level and municipal-level food security as well as to encourage local exchange and overall rural development. These early initiatives put food sovereignty principles into practice, but the term itself was rarely used and thus was little known outside LVC circles.

In 2001 the World Forum on Food Sovereignty (WFFS), the first major international meeting focused specifically on food sovereignty since the introduction of the concept by LVC in 1996, was held in Havana, Cuba. In the context of Nicaragua, the WFFS was considered to be a turning point by bolstering knowledge about food sovereignty. Nicaraguan organisations belonging to LVC, as well as those from other Nicaraguan CSOs, participated in the event as delegates and brought back a more articulated concept of food sovereignty. The *Declaration of the World Forum on Food Sovereignty* and other documents about food sovereignty produced by LVC and other movements and organisations began to be circulated among different groups in Nicaragua and were influential in shaping the opinions of members of other CSOs, some of which were directly involved in national-level initiatives to promote food security in Nicaragua.[32] While the concept of food sovereignty at this time was certainly not widespread, it was gradually becoming more known among farmer organisations and CSOs working on food security and agricultural production issues in Nicaragua.

Initial attempts at national food security policy in Nicaragua: 1998–2001

In response to commitments made at the 1996 WFS to adopt policies to guarantee the right to food and strengthen national food security,[33] and parallel to early food sovereignty mobilisation, there was a national concerted effort to pursue food security policy in Nicaragua in the 1997–2001 period. The deteriorating food security situation faced by rural areas, primarily in the dry zones of Nicaragua and mainly affecting children,[34] was only complicated by the effects of Hurricane Mitch in 1998 and the Coffee Crisis in the early 2000s. This led to a push for national food security policy to support Nicaragua's constitutional amendment guaranteeing the right to food.

Two broad food security initiatives were pursued. The first of these was led by Sandinista deputies to the National Assembly who, in coordination with the Food and Agriculture Organization (FAO) (which had worked with the

Revolutionary government in the 1980s on their food security agenda) and other food security stakeholders in Nicaragua,[35] introduced draft legislation for a food security law in 1997. After the initial proposal for the law failed to advance through the legislative process, a second, strengthened proposal, entitled the Food and Nutritional Security Law (FNS Law), was drafted and introduced to the National Assembly in 2001. The drafting of the FNS Law was influenced by another, simultaneous initiative. In 2000, via presidential decree, the Alemán government introduced two new food security institutions,[36] the National Committee of Food and Nutritional Sovereignty and Security (CONASAN) and the Technical Committee of Food and Nutritional Security (COTESAN), as well as a National Food and Nutritional Security Policy (PNSAN).[37] The Alemán government initially assigned a government agency, the Secretariat of Social Action, with the responsibility for the formulation of the policy, which was also undertaken with the participation of various government institutions, international agencies such as the FAO and the World Food Programme (WFP), and CSOs working on food and nutritional security.[38] The PNSAN's accompanying Plan of Action was released in 2001.

Despite the efforts of promoters of the FNS Law and the intention of the PNSAN to bolster food security in Nicaragua, both failed in their own ways to effect change. While the updated proposal for the FNS Law was successfully introduced to the National Assembly in January of 2001, it lay dormant in the legislative process for years.[39] The main obstacle facing the bill was a 'lack of political will', as the food security issue was reportedly not prioritised by the predominantly liberal government.[40] On the other hand, the PNSAN and its accompanying Plan of Action were never effectively implemented.[41] A former FAO official explained very candidly that, despite attempts to work with the policy, 'it was more the face of propaganda' and 'never managed to achieve its purpose'.[42] It was not until several years later that efforts to pursue a food security law were revived, only this time the initiative was spearheaded by Nicaraguan rural organisations belonging to LVC seeking to introduce a national food security policy that explicitly included the concept of food sovereignty.

Mobilising for food sovereignty policy in Nicaragua

LVC Central America began to discuss pursuing food sovereignty policy legislation at the national level throughout the region in the 1998–2000 period; this initiative was deepened in the years following the 2001 WFFS. In the context of Nicaragua the call for national food sovereignty policy by Nicaraguan LVC organisations was strengthened with the formation of the Interest Group for Food and Nutritional Sovereignty and Security (GISSAN), founded in April 2004 by a group of organisations belonging to LVC, as well as by Nicaraguan and international CSOs working on food security, rural development, sustainable agriculture and public policy.[43] UNAPA and the Nicaraguan Soy Association (SOYNICA), a Nicaraguan CSO active in food security and nutrition, played fundamental roles in the establishment of the interest group and its subsequent coordination. GISSAN's stated mission at this time emphasised a commitment to the promotion of 'food sovereignty through the impact of public policies and people' and visualised itself as a 'permanent forum, recognized nationally and internationally as a supporter in the struggle for Food Sovereignty, in which proposals for SSAN [Food and Nutritional Sovereignty and Security] and indicators of advances, support, lobbying, and impact at the municipal, national, and international levels are discussed, formulated, and promoted'.[44]

The creation of GISSAN can be seen as an important step towards the consolidation of the food sovereignty movement in Nicaragua. However, initially most members of GISSAN – with the exception of Nicaraguan LVC members, Oxfam Belgium, Oxfam Intermon and SOYNICA – were sceptical of food sovereignty, as the term was unknown to some and for others had a political focus that was not seen as addressing the nutritional concerns many of these organisations focused on through their work.[45] As 'food security' was a far more familiar term, member organisations debated whether to include 'security' in the name of the group or simply use 'sovereignty'.[46] What resulted was the term 'food and nutritional sovereignty and security', as reflected in the name of GISSAN, since a consensus by the members on solely using the term 'sovereignty' could not be reached.

Parallel to the founding of GISSAN, in 2004–05 the ATC and UNAPA decided to move forward with the formulation of a food sovereignty law in Nicaragua. A proposal for a Law of Food and Nutritional Sovereignty and Security (FNSS Draft Law) was drafted by leaders of two farmer organisations belonging to MAF and its content reflected the ideological perspectives of Nicaraguan LVC organisations. The content of this first draft of the FNSS law included elements of the FNS Law and the PNSAN, while introducing the concept of food sovereignty and including key food sovereignty principles. The FNSS Draft Law was also circulated among GISSAN's membership and shared with a key FSLN National Assembly deputy, Dora Zeledón, leader of the initiative for the FNS Law and a UNAPA ally, who reportedly expressed her support for the new proposal.

Not long after the draft was produced and circulated, GISSAN was approached by an aide to high-level FSLN National Assembly Deputy Walmaro Gutiérrez, to see if the group would be interested in drafting a food security law. As some of the groups belonging to GISSAN had been active in promoting the FNS Law in the 1998–2001 period, members contemplated whether they should revive that law but it was finally decided that GISSAN would update the 'famous law written by UNAPA'.[47] The process of updating the FNSS Draft Law was undertaken in 2006 and consisted of broad consultations in municipalities throughout Nicaragua. The consultations were sponsored by organisations and networks belonging to GISSAN (especially GPAE), many of which had strong links to local communities in rural areas through their grassroots-based activities and historical relationships. The consultation process consisted of local events at which the FNSS Draft Law was presented and participating citizens shared their comments and suggestions, which were recorded by GISSAN members. Following the consultation process, representatives of GISSAN met to update the FNSS Draft Law, incorporating the feedback gathered from the consultations.

Several other important developments occurred parallel to the growth of the food sovereignty movement and the elaboration of the FNSS Draft Law that further redirected attention towards promoting legislation to support food security and the right to food, but did not include a focus on food sovereignty. At the national level, two FAO-supported initiatives were undertaken with universities and were strategic in terms of moving the food security dialogue forward: a partnership between the FAO and several universities was established in 2005–06 to offer a graduate-level programme in food and nutritional security specifically tailored to food security professionals and technicians, which was reported to raise awareness of the importance of right to food legislation,[48] and the Inter-University Council for Food and Nutritional Security (CIUSAN) was founded by four prominent national universities in 2004. At the regional level, Guatemala passed its National Food and Nutritional Security System Law in April 2005, thus becoming the first nation in Central America to pass national food security legislation. A second regional development was the FAO-supported creation of the Hunger-Free Latin America and Caribbean Initiative at the Latin American Summit on Chronic Hunger, held in Guatemala in September 2005, at which Latin American heads of state committed to eradicating hunger in the region by 2025; the primary means of this would be through the development of national legal frameworks to ensure the right to food.[49] Last but not least, the debate over the Dominican Republic–Central American Free Trade Agreement (DR–CAFTA) characterised the early 2000s and intensified leading up to the adoption of the agreement, which entered into force in Nicaragua in 2006.

GISSAN's draft bill enters the National Assembly

In September 2006, within this context of increasing national and regional focus on the issue of food security and the intensifying call for national policy to support the right to food, the updated GISSAN version of the FNSS Draft Law was quietly introduced by Deputy Gutiérrez to the National Assembly. As per the legislative process the proposal was sent to a committee, in this case the Special Committee to Monitor Poverty Reduction, of which Gutiérrez was president, to be reviewed. The committee gave the proposal a favourable *dictamen* (opinion) in early October and requested that it be included immediately on the National Assembly plenary agenda for debate.[50]

Meanwhile, GISSAN focused its attention on campaigns to raise public awareness about the concept of food sovereignty and the proposal for the law. The reason for this, one GISSAN representative explained, was 'because the more people who knew the word, the more people would understand it, and more people would use the word rather than security'.[51] However, as the representative further explained, there was continued resistance to adopting the term both within and outside GISSAN:

> There was much opposition against the word – technical people in ministries, field technicians of civil organisations who barely came to understand food security and nutrition [said to us] 'you are giving us another concept and we are completely confused'. Also in GISSAN, in our strategic plan, we only used 'food sovereignty', saying that if there is food sovereignty, food security is there. [Some GISSAN members] did not want to accept this.[52]

Just a month after the FNSS Draft Law received a favourable *dictamen*, a major event took place that had profound influence on the political environment:

Daniel Ortega's presidential election victory in the November 2006 general election brought the FSLN back to control the executive branch of government after 16 years of liberal leadership.[53] The incoming Sandinista government prioritised food security and recognised food sovereignty, at least in rhetoric. Within the first few months of assuming power, the government had created the Food Security and Sovereignty Council (CSSA) to replace the CONASAN (see above). The appointed head of CSSA had strong links to the Revolution and was the director of a GISSAN member organisation. He was also instrumental in launching a second government initiative, the 'Zero Hunger' programme, modelled on the BPA experience promoted by the early food sovereignty movement in the 1990s.[54] A law, from the perspective of the government, would only serve to strengthen the legitimacy of such programmes.

With the new government's keen interest in approving the pursuit of food security legislation, the head of the Sandinista *bancada* (political party) in the National Assembly asked Dora Zeledón to strengthen the FNSS Draft Law in early January 2007 and address cited weaknesses mostly pertaining to the legal technique of law.[55] She reportedly invited the FAO, which was seen by deputies to be an expert on the subject of food security and right-to-food legislation, to provide technical support and also consulted other government actors and CSOs over the months that followed.[56] An important workshop was sponsored by the NGO World Vision in early June 2007 that brought different stakeholders in the area of food security together, including representatives of the FAO, WFP, GISSAN (including LVC organisations), representatives from the High Council of Private Enterprise (COSEP),[57] and National Assembly deputies, to discuss the FNSS Draft Law. Several technical critiques of the *dictamen* were made at the event, namely by FAO consultants and/or officials, which cited the need to polish the legal technique of the bill, to incorporate a more elaborate institutional system to implement the law similar to what had been done in Guatemala), and finally to include sanctions for non-compliance with the law.[58] An agreement was reached between GISSAN, COSEP and Dora Zeledón in terms of how the law would be revised. Furthermore, FAO's Special Programme for Food Security (FAO-PESA) committed to providing technical support for the final negotiation of the law and funding to facilitate the process of its approval.

June 2007: the breakdown in the approval of the law by the National Assembly

Before revisions could be made to the FNSS Draft Law based on the agreement forged between stakeholders at the World Vision event, the bill was added to the National Assembly plenary agenda and introduced onto the floor for debate on 12 June 2007, at which time it was approved in general by 80 votes.[59] The debate on the law, and its 'approval in the particular',[60] continued the next day with the first four articles of the law being approved before the debate turned to Article 5, which led to intense discussion among the deputies.

Subsections of Article 5 represented key issues for food sovereignty advocates, including protection for small and medium-sized production and producers and controls on food imports, including the prohibition of food aid containing

GMOs, with the latter being the most controversial point in the National Assembly debate. The deputies were divided. Left-wing *bancadas*, the FSLN and the Sandinista Renovation Movement (MRS), defended the content of Article 5 and raised concerns about the dangers of GMOs. Liberal *bancadas* opposed elements of Article 5, especially the subsection on food import controls and the provision on GMOs, and some strongly favoured suspending the debate. After it became clear that there were significant factors that needed to be resolved and negotiated, and that the only way to do this was through motions, the debate over the law was suspended until 20 June, with the expectation that the different *bancadas* within the National Assembly would reach agreements through the introduction of motions so that the law could be approved in the next round of scheduled debates.[61]

One of the major forces behind the suspension of the debate was COSEP. When COSEP realised that the law was being debated in the plenary and was approved both in general and through Article 4, it contacted the *bancada* of the Constitutionalist Liberal Party (PLC), an opposition party closely aligned with the private sector, and urged the halting of the debate on the law. This was confirmed by a COSEP representative who explained that the law 'did not represent the foundations of food security' and, furthermore, the law was 'far from contributing to food sovereignty and food independence' because of the risks it presented to the private sector.[62] To stop the debate, COSEP immediately explained its concerns to its PLC allies in the National Assembly, who acted to suspend the approval process.

The major problems with the FNSS Draft Law, according to COSEP, were that limited public consultation on the law had been carried out;[63] the law did not guarantee equal treatment, as it proposed preferential treatment for small and medium producers at the expense of large-scale producers; the law potentially allowed the state to intervene in both national marketing and distribution chains as well as international trade; and, related to the previous point on trade, the language concerning the regulation of GMOs, which called for the 'strict control that permits the entry of harmless foods into the country for consumption, not permitting the receiving of food aid that contains genetically-modified material', was ambiguous.[64] Finally, according to a policy insider, the very concept of food sovereignty was a critical point of contention for the private sector, which argued through the PLC *bancada* that it was not going to approve the law if the definition of food sovereignty remained as it was written in the *dictamen*.

From the point of view of other stakeholders the opposition of the private sector had to do with protecting that sector's interests. Pointing out that the members of COSEP were large-scale businesses, it was suggested that some – but not all – of the members of COSEP opposed the bill because of their economic interests in the agrochemical, basic grains, and genetically-modified seed and food trade. These sentiments were echoed by others, citing not only the private sector's support for free trade and the importation of transgenic products, but also the FSLN government's defence of the sector. As several representatives of farmer organisations and CSOs alleged, FSLN leaders were reported to have links to commercial interests in GMOs, thus National Assembly deputies were hesitant to approach the issue of transgenics for fear of retaliation.[65] FSLN support for the private sector was explained by some to be part of the party's broader strategy in terms of retaining power and it was further explained that the FSLN suffered on account of having marginalised the private sector during the Revolution and prioritising peasant interests. Having regained power, and with some high-ranking cadres having strong links to agribusiness, the FSLN was interested in maintaining a positive relationship with the private sector, at times at the expense of small and medium farmers.

The private sector's opposition to the law was further explained as being attributable to a different understanding of how to achieve food security. As the representative of COSEP explained, 'the true basis of food sovereignty has always existed. It is the right of peoples to determine their own policies and states always have institutions that are responsible for that. So for me it's not new.'[66] From COSEP's perspective, it was not food sovereignty that was needed to achieve food security but rather that of 'food independence', which the COSEP representative described in the following way:

> [Food independence] is the ability of a people or a state to be self-sufficient in its own food production so that imports are not necessary and instead if production is surplus may have export opportunities. That concept does not limit imports, ie if the country is self-sufficient, it does not require imports. But this concept does not mean a ban on imports, rather imports with domestic production come to create an over-availability, but the basis of food security should be self-sufficiency of the state in food production. Obviously, if this self-sufficiency is not achieved, one turns to imports, but when you effectively produce enough, imports come to give additional security because in addition to your own production that produces food for the country, imports come to create an additional source that has many advantages.[67]

COSEP's notion of food independence shared some compatibility with food sovereignty in terms of a focus on working towards food self-sufficiency. However, the privileging of open markets and opposition to forms of state regulation and market protections, such as the regulation of products containing GMOs, were clear and important differences.

Discussion and concluding thoughts

The story of Law 693 highlights the important question of the potential for food sovereignty to be meaningfully incorporated into national public policies, given the controversial nature of its agenda. Although certain conditions created opportunities for food sovereignty to be included in the law, the concept and its principles faced important obstacles that began to emerge with the formation of GISSAN, strengthened during the formulation of the FNSS Draft Law, and deepened with the breakdown in the approval process. Above all, despite the efforts of the movement to promote the concept to the broader public, food sovereignty was neither clearly understood nor widely appropriated. It is here that the three challenges identified earlier – conceptual ambiguity, competing approaches to food security, and the contradictions posed by the state – become visible and interact in such a way as to create conditions that jeopardised the successful institutionalisation of food sovereignty as it was initially envisioned in the FNSS Draft Law.

First, the food sovereignty concept was both relatively unknown in Nicaragua and reportedly confusing. One reason for this specific to the Nicaraguan context was the historical significance of sovereignty as employed during Sandino's struggle and the Revolution, which was more aligned with political sovereignty and autonomy than with the framework of food sovereignty. Furthermore, the meaning of food sovereignty was also confusing when compared with food security. This was perhaps best captured in the term 'food sovereignty and security', which was perplexing for many food sovereignty supporters because, according to the established discourse, food sovereignty is a means to achieving genuine food security.[68] The term 'food sovereignty and security' runs the considerable risk of not making this distinction clear and also suggests that food sovereignty was undergoing a process of reinterpretation by actors as they engaged in a process of attaching meaning to the term. Comments made by the COSEP representative, that food sovereignty had 'always existed', further suggested alternative interpretations of the concept to that of the broader movement. All these factors point to the lack of coherence in the food sovereignty discourse in Nicaragua and a lack of shared understanding as to what the concept meant, both of which undermine the force of food sovereignty and, by extension, unity behind policy initiatives seeking to include its conceptual framework.

Given the lack of discursive force of the food sovereignty approach, it faced significant challenge when the FNSS Draft Law entered the formal policy-making process – a critical site of political struggle – and was put to the test against competing approaches to food security advocated by powerful interests. First there was the more technical approach of the FAO. Considered to be an authority on food security and right-to-food legislation, the FAO was able to exert influence on the process at the behest of the state. The FAO approach has to be understood within the context of its broader campaign for national framework laws (eg a central component of the Hunger-Free Latin American initiative) to support the right to food, which are largely based on overarching guidelines that take a more top-down approach to policy.[69] Then there was the private sector, which had considerable political weight in policy making (evidenced by the remarkable speed with which it was able to stop the approval process of the law), and its alternative vision of 'food independence', which emphasised the importance of economic openness. Lastly was the position of the state under the leadership of the FSLN, which reflected ideas and beliefs that were arguably more in line with food self-sufficiency and production, reminiscent of the Revolution, and sought to introduce new mechanisms, like the Zero Hunger Programme, to accomplish these goals. In analysing the dynamics of power between the different 'owners' of these policy frames, it is not so simple as to say that the interests of competing frames were opposed to those of the food sovereignty movement – there was clear overlap between aspects of their individual agendas as well as divergence. The key point that rather needs to be made here concerns the capacity of food sovereignty to present a convincing and viable alternative to approaches advocated by more powerful stakeholders, which was not evident at this point in the policy-making process.

Finally, regarding the state, the FSLN government was one of the most crucial players in this story and demonstrated clear political will to support food security policy initiatives, with FSLN deputies creating political opportunities for the FNSS Draft Law to advance through the National Assembly. However, while state actors began to create space for policy change, this was largely inadequate for the kinds of the reforms envisioned under food sovereignty. Several factors limited the state' ability to create the conditions necessary to foster food sovereignty. DR-CAFTA, for example, to which Nicaragua was party, prohibited the institutionalising of provisions to protect special groups (ie small and medium producers) or to prohibit the entry of GMOs, and there was little indication that the state planned to renege on its obligations or challenge the private sector, with which it was reportedly interested in maintaining positive relations. Furthermore, while it is important to note that the Sandinista government at this time was experimenting with new strategies to support participatory governance, its decision to invite the FAO to provide technical assistance and expertise, given the food security policy frame of the FAO described above and the latter's relative power, compromised the spirit of the participatory and democratic formulation of food and agriculture policy envisioned by the food sovereignty approach.[70] Echoing Clark, the state was indeed both a solution in some instances and a problem in others.[71]

In closing, the case of Nicaragua provides a rich empirical example of the complexities of food sovereignty policy making. In addition to underscoring the importance of shared interpretations of food sovereignty for building discursive force, it also highlights the hurdles faced by food sovereignty in asserting its legitimacy in the face of stronger and often historically entrenched policy alternatives. Finally, this study provides an important example of how 'leftist', 'post-neoliberal' states that proclaim their support for food sovereignty struggle to negotiate their relationship with social and economic actors, trying to avoid pitfalls of the past and also trying to build their own legitimacy, arguably at the cost of those they claim to serve.

Acknowledgments

The author expresses her appreciation to Fidaa Shehada, Falguni Guharay, Nils McCune, three anonymous reviewers, and especially Christina Schiavoni for their thoughtful insights and suggestions on this contribution.

Notes

1. Wittman et al., "Seeing like a Peasant," 2. Since its introduction by LVC, the concept has evolved and become more nuanced. This is easily traced via declarations from successive international meetings addressing food sovereignty, at which the concept has been collectively deliberated and rearticulated. Despite variances, most definitions of food sovereignty share common elements.
2. Edelman, 'Food Sovereignty."

3. Edelman, 'Transnational Organizing in Agrarian Central America"; Edelman, "Food Sovereignty"; and Martinez-Torres and Rosset, "La Vía Campesina."

4. See Desmarais, *Globalization and the Power of Peasants*; and Rosset, "Food Sovereignty." For a description of the productionist approach, see Lang and Heasman, *Food Wars*.

5. See, for example, McMichael, "Global Development and the Corporate Food Regime."

6. La Vía Campesina, "The Right to Produce."

7. See Araújo, "The Promise and Challenges of Food Sovereignty Policies"; Araújo and Godek, "Can Food Sovereignty Laws make the World more Democratic?"; Beauregard, "Food Policy for People"; and Wittman et al., "Seeing like a Peasant."

8. See Araújo and Godek, "Can Food Sovereignty Laws make the World more Democratic?"; Beauregard, "Food Policy for People"; Wittman et al., "Seeing like a Peasant"; and Giunta, "Food Sovereignty in Ecuador."

9. For comprehensive discussions of the food sovereignty framework and its evolution, see Desmarais, *Globalization and the Power of Peasants*; Pimbert, *Towards Food Sovereignty*; and Windfuhr and Jonsén, *Food Sovereignty*.

10. Lang and Heasman, *Food Wars*.

11. Mooney and Hunt, "Food Security."

12. Gamson, "Constructing Social Protest."

13. Candel et al., "Disentangling the Consensus Frame of Food Security."

14. Moony and Hunt, "Food Security."

15. Candel et al., "Disentangling the Consensus Frame of Food Security."

16. See, for example, Fairclough, *Discourse and Social Change*.

17. Patel, 'What does Food Sovereignty look Like?"

18. Ibid; and Wittman and Desmarais, "Farmers, Foodies and First Nations." See also Drolet et al., "Food Security in Nicaragua." Drolet et al.'s unpublished study tested an FAO policy analysis framework and placed very little emphasis on the food sovereignty dimension of Law 693; however, they did find that key Nicaraguan stakeholders – some of whom participated in making Law 693 – understood food sovereignty in significantly different ways.

19. Boyer, "Food Security," 333; Claeys, 'The Creation of New Rights"; Mesner, "The Territory of Self-determination"; and Shiavoni, "Competing Sovereignties."

20. Boyer, "Food Security," 333.

21. Ibid., 334.

22. Clark, "Food Sovereignty," 7. See also Peschard, "Farmers' Rights and Food Sovereignty."

23. McKay et al., "The 'State' of Food Sovereignty in Latin America."

24. Shiavoni, "Competing Sovereignties," 3. See also Giunta, "Food Sovereignty in Ecuador" for a discussion of how such relations are being renegotiated in Ecuador; and Trauger, "Toward a Political Geography of Food Sovereignty."

25. See Clark, "Food Sovereignty."

26. In Nicaragua, the concept of 'sovereignty' is historically significant and can be traced back to Augusto César Sandino's popular nationalist movement in the 1920s and 1930s. Thanks to the influence of Sandino in the Revolution, the concept was an intrinsic part of revolutionary rhetoric.

27. See, for example, Austin et al., "The Role of the Revolutionary State"; Biondi-Morra, *Hungry Dreams*.

28. For a rich description of PCAC, see Holt-Giménez, *Campesino a Campesino*.

29. For a rich description of PCAC, see E. Holt-Giménez, Campesino a Campesino: Voces de Latinoamérica – Movimiento Campesino a Campesino para la Agricultura Sostenible, Managua: SIMAS, 2008.

30. Interview with MAF representative, July 27, 2012.

31. The ATC was instrumental in the establishment of both UNAPA, which represents small producers, and MAF, which is comprised of six national agriculture and forestry organisations, including the ATC and UNAPA. In addition to the ATC, UNAPA and MAF became members of LVC. UNAG left LVC in 1996–97, although it remained the home of PCAC-Nicaragua.

32. Interview with GISSAN representative, August 24, 2011.

33. See FAO, *Rome Declaration on World Food Security*.

34. D. Zeledón, "Proceso de la Ley de SSAN en Nicaragua," National Assembly of Nicaragua, Managua, n.d, 1.

35. Including United Nations agencies, international institutions, international cooperation agencies, government agencies, NGOs and universities.

36. The Alemán government, as well as the governments that preceded and followed it (the Chamorro and Bolaños administrations), were neoliberal in their orientation.

37. Alemán Lacayo, "De Creación de la Comisión Nacional."

38. Lorio Castillo, *Avances en la aplicación de la Ley de Soberanía*, 8; and FAO, "Estado de la Seguridad Alimentaria."

39. See National Assembly of Nicaragua for a chronology of the bill once it was introduced. Accessed April 4, 2014, http://www.asamblea.gob.ni/trabajo-legislativo/agenda-legislativa/ultimas-iniciativas-dictaminadas.

40. "Interview with Dora Zeledón – April 2002," http://www.rdfs.net/news/interviews/zeledon-apr2002_en. htm.
41. For more in-depth discussion of the PNSAN's limitations, see Lorio Castillo, *Avances en la aplicación de la Ley de Soberanía*; and Sahley et al., *The Governance Dimensions*.
42. Interview, July 6, 2012.
43. A MAF representative stated that the impetus behind the formation of GISSAN came from LVC to promote food sovereignty through the establishment of links between farmer organisations and CSOs. Personal communication, February 7, 2013. At one point in the mid-2000s, GISSAN membership peaked at 70-some organisations. GISSAN's membership included organisations that were also part of other organisational networks and coalitions tackling issues that were expressly part of the food sovereignty framework and included in GISSAN's platform, namely the fight against GMOs, water privatisation, and the negotiation of what would become the Dominican Republic–Central American Free Trade Agreement.
44. The mission and vision statements quoted here were retrieved from GISSAN's former website, http://gissannicaragua.org, in April 2010. This website is no longer active.
45. Personal communication with former GISSAN/UNAPA representative, October 24, 2014.
46. Interview with GISSAN representative, August 24, 2011.
47. Ibid.
48. Interview with Food Policy Consultant, June 27, 2012.
49. See Hunger-Free Latin America and Caribbean website, at http://www.rlc.fao.org/en/initiative/the-initiative, accessed 4 April 2014.
50. See http://www.asamblea.gob.ni/trabajo-legislativo/agenda-legislativa/ultimas-leyes-aprobadas/ for a copy of the letter sent by the president of the committee, Deputy Walmaro Gutiérrez, to the First Secretary of the National Assembly, dated October 5, 2005.
51. Interview with GISSAN representative, August 24, 2011.
52. Ibid.
53. The National Assembly, however, was still controlled by the Liberal Constitutional Party (PLC).
54. 'Zero Hunger' provides rural women with seeds, small livestock, and technical assistance to encourage food production and greater family and community food security. CSSA was short-lived. After six months the director was reappointed to lead the Zero Hunger programme.
55. Interviews with National Assembly deputies, August 21 and 27, 2011.
56. Interview with National Assembly deputy, August 21, 2011; Zeledón, "Proceso de la Ley," 2.
57. COSEP is the principle advocacy organisation for the private sector in Nicaragua and plays a strong role in policy formation, with its representatives working not only on national policies but also on international policies, including DR-CAFTA.
58. Interview with former FAO official, March 7, 2012.
59. There are 92 deputies in the National Assembly, thus the proposed law received strong support from the legislature.
60. Following the approval of the law in general, the law is then 'approved in the particular' by the deputies, meaning that each article of the law is reviewed, debated and voted on. Once a law is approved in general, no changes can be made to its content except through motions.
61. For the transcripts of the National Assembly debates on the law, see http://www.asamblea.gob.ni/trabajo-legislativo/diario-de-debates/.
62. Interview with former FAO official, March 7, 2012; and interview with COSEP representative, June 26, 2012.
63. A step typically carried out by the National Assembly before the passing of a law in which different societal actors are consulted as to their position on proposed legislation.
64. From the viewpoint of the private sector, it was unclear, first, what 'strict control' referred to and, second, how food safety would be assessed without having clear means by which to measure or analyse the risk posed by products containing genetically modified material.
65. Interview with Campesino a Campesino representative, June 6, 2012; and interview with Centro Humboldt representative, March 11, 2013. It was explained that many Sandinista leaders had become businessmen in the 1990s and had investments or strong ties to powerful agribusinesses.
66. Interview with COSEP representative, June 26, 2012.
67. Ibid.
68. See, for example, Murphy, "Expanding the Possibilities for a Future Free of Hunger."
69. For example, Committee on Economic, Social and Cultural Rights (CESCR), "Substantive Issues Arising in the Implementation of the International Covenant on Economic, Social, and Cultural Rights: General Comment 12," Twentieth Session, Geneva, April 26–14 May 1999, subsection 29, at http://www.unhchr.ch/tbs/doc.nsf/0/3d02758c707031d58025677f003b73b9; and FAO, *The Right to Food in Practice*.
70. See Müller, "The Temptation of Nitrogen."
71. Clark, "Food Sovereignty."

Bibliography

Alemán Lacayo, Arnoldo. "De Creación de la Comisión Nacional de Seguridad Alimentaria y Nutricional, Decreto No. 40-2000." *La Gaceta: Diario Oficial*, no. 92 (2000). http://bibliotecageneral.enriquebolanos. org/gacetas_siglo_XXI/G-2000-05-17.pdf.

Araujo, Saulo. "The Promise and Challenges of Food Sovereignty Policies in Latin America." *Yale Human Rights and Development Law Journal* 13, no. 2 (2010): 99–112.

Araujo, Saulo, and Wendy Godek. "Can Food Sovereignty Laws make the World more Democratic? The Case of Nicaragua's Food Sovereignty Law." In *Access to Food in the New Millennium*, edited by Lea Brilmayer, Nadia Lambek, and Adrienna Wong. New York: Springer, forthcoming.

Austin, J., J. Fox, and W. Kruger, "The Role of the Revolutionary State in the Nicaraguan Food System." *World Development* 13, no. 1 (1985): 15–40.

Biondi-Morra, Brizio N. *Hungry Dreams: The Failure of Food Policy in Revolutionary Nicaragua, 1979–1990*. Ithaca, NY: Cornell University Press, 1993.

Beauregard, Sadie. "Food Policy for People: Incorporating Food Sovereignty Principles into State Governance: Case Studies of Venezuela, Mali, Ecuador, and Bolivia." 2009. http://www.oxy.edu/sites/default/files/ assets/UEP/Comps/2009/Beauregard%20Food%20Policy%20for%20People.pdf.

Boyer, Jefferson. "Food Security, Food Sovereignty, and Local Challenges for Agrarian Movements: The Honduras Case." *Journal of Peasant Studies* 37, no. 2 (2010): 319–351.

Candel, J. J. L., G. E. Breeman, S. J. Stiller, and C. J. A. M. Termeer, "Disentangling the Consensus Frame of Food Security: The Case of the EU Common Agricultural Policy Reform Debate." *Food Policy* 44 (2014): 47–58.

Claeys, Priscilla. "The Creation of New Rights by the Food Sovereignty Movement: The Challenge of Institutionalizing Subversion." *Sociology* 46, no. 5 (2012): 844–860.

Clark, P. "Food Sovereignty, Post-neoliberalism, Campesino Organizations and the State in Ecuador." Paper presented at the 'Food Sovereignty: A Critical Dialogue' international conference, Yale University, September 14–15, 2013. http://www.yale.edu/agrarianstudies/foodsovereignty/pprs/34_Clark_2013.pdf.

Desmarais, Annette Aurélie. *Globalization and the Power of Peasants*. Halifax/London: Fernwood Publishing/ Pluto Press, 2007.

Drolet, Adam, Mark Fitzsimmons, Megan Montgomery, and Alyson Platzer. *Food Security in Nicaragua: Policy Analysis and Preparedness Pilot Study – Final Report*. Washington, DC: FAO/Elliot School of International Affairs, George Washington University, May 13, 2011.

Edelman, M. "Food Sovereignty: Forgotten Genealogies and Future Regulatory Challenges." Paper presented at the 'Food Sovereignty: A Critical Dialogue' international conference, Yale University, September 14–15, 2013. http://www.yale.edu/agrarianstudies/foodsovereignty/pprs/72_Edelman_2013.pdf.

Edelman, M. "Transnational Organizing in Agrarian Central America: Histories, Challenges, Prospects." *Journal of Agrarian Change* 8, nos. 2–3 (2008): 227–259.

Food and Agriculture Organization (FAO). "Estado de la Seguridad Alimentaria y Nutricional en Nicaragua." Paper presented at the workshop 'Hacia la elaboración de una estrategia de asistencia técnica de la FAO en apoyo a la implementación de la Iniciativa América Latina y el Caribe Sin Hambre', Guatemala City, October 18–19, 2006. http://www.rlc.fao.org/fileadmin/templates/iniciativa/content/pdf/publicaciones/asisten cia-programas-san/sanic.pdf.

FAO. *The Right to Food in Practice: Implementation at the National Level*. Rome: FAO, 2006. http://www. fao.org/docrep/016/ah189e/ah189e.pdf.

FAO. *Rome Declaration on World Food Security and the World Food Summit Plan of Action*. World Food Summit, Rome, November 13–17, 1996. http://www.fao.org/docrep/003/w3613e/w3613e00.htm.

Fairclough, N. *Discourse and Social Change*. Cambridge: Polity, 1992.

Gamson, W. A. "Constructing Social Protest." In *Social Movements and Culture*, edited by H. Johnston and B. Klandermans, 85–106. Minneapolis: University of Minnesota Press, 1995.

Giunta, I. "Food Sovereignty in Ecuador: Peasant Struggles and the Challenge of Institutionalization." *Journal of Peasant Studies* 41, no. 6 (2014): 1201–1224.

Holt-Giménez, E. *Campesino a Campesino: Voces de Latinoamérica – Movimiento Campesino a Campesino para la Agricultura Sostenible*. Managua: SIMAS, 2008.

Lang, T, and M Heasman. *Food Wars: The Global Battle for Mouths, Minds, and Markets*. London: Earthscan, 2004.

La Vía Campesina. "The Right to Produce and Access to Land." Paper presented at the World Food Summit, Rome, November 13–17. 1996.

Lorio Castillo, Margarita. *Avances en la aplicación de la Ley de Soberanía y Seguridad Alimentaria y Nutricional*. Colección Diálogo Social 7. Managua: Instituto de Estudios Estratégicos en Políticas Públicas (IEEPP), 2011.

McKay, B., R. Nehring, and M. Walsh-Dilley, "The 'State' of Food Sovereignty in Latin America: Political Projects and Alternative Pathways in Venezuela, Ecuador and Bolivia." *Journal of Peasant Studies* 41, no. 6 (2014): 1175–1200.

McMichael, Philip. "Global Development and the Corporate Food Regime." In *New Directions in the Sociology of Global Development*, edited by Frederick H. Buttel and Philip McMichael, 265–299. Bingley, UK: Emerald Group Publishing, 2005.

Martinez-Torres, M. E. and P. M. Rosset. "La Vía Campesina: The Birth and Evolution of a Transnational Movement." *Journal of Peasant Studies* 37, no. 1 (2010): 149–175.

Mesner, M. "The Territory of Self-determination: Social Reproduction, Agro-ecology, and the Role of the State." In *Food Sovereignty: Global and Local Change in the New Politics of Food*, edited by P. Andrée, J. Ayres, M. J. Bosia, and M.-J. Massicotte. Toronto: University of Toronto Press, 2014.

Mooney, P. H., and S. A. Hunt. "Food Security: The Elaboration of Contested Claims to a Contested Frame." *Rural Sociology* 74, no. 4 (2009): 469–497.

Müller, B. "The Temptation of Nitrogen: FAO Guidance for Food Sovereignty in Nicaragua." Paper presented at the 'Food Sovereignty: A Critical Dialogue' international conference, Yale University, September 14–15, 2013. http://www.yale.edu/agrarianstudies/foodsovereignty/pprs/33_Muller_2013.pdf.

Murphy, S. "Expanding the Possibilities for a Future Free of Hunger." *Dialogues in Human Geography* 4 (2014): 225–228.

Patel, R. "What does Food Sovereignty look Like?" *Journal of Peasant Studies* 36, no. 3 (2009): 663–706.

Peschard, K. "Farmers' Rights and Food Sovereignty: Critical Insights from India." *Journal of Peasant Studies* 41, no. 6 (2014): 1085–1108.

Pimbert, Michel. *Towards Food Sovereignty: Reclaiming Autonomous Food Systems*. London: International Institute for Environment and Development, 2009.

Rosset, P. "Food Sovereignty: Global Rallying Cry of Farmer Movements." *Food First Backgrounder* 9, no. 4 (2003): 1–4.

Sahley, Caroline, Benjamin Crosby, David Nelson, and Lane Vanderslice. *The Governance Dimensions of Food Security in Nicaragua*. Accessed January 5, 2015. http://pdf.usaid.gov/pdf_docs/PNADE106.pdf.

Shiavoni, C. "Competing Sovereignties, Contested Processes: The Politics of Food Sovereignty." Master's thesis, International Institute of Social Studies, 2013. http://thesis.eur.nl/pub/15217.

Suppan, Steve. "Challenges for Food Sovereignty." *Fletcher Forum of World Affairs* 32, no. 1 (2008): 111–123.

Trauger, A. "Toward a Political Geography of Food Sovereignty: Transforming Territory, Exchange and Power in the Liberal Sovereign State." *Journal of Peasant Studies* 41, no. 6 (2014): 1131–1152.

Windfuhr, Michael, and Jennie Jonsén. *Food Sovereignty: Towards Democracy in Localized Food Systems*. Rugby, UK: ITDG Publishing, 2005.

Wittman, Hannah, and Annette Desmarais. "Farmers, Foodies and First Nations: Getting to Food Sovereignty in Canada." Paper presented at the 'Food Sovereignty: A Critical Dialogue' international conference, Yale University, September 14–15, 2013. http://www.yale.edu/agrarianstudies/foodsovereignty/pprs/3_Desmarais_Wittman_2013.pdf.

Wittman, Hannah Kay, Annette Aurelie Desmarais, and Nettie Wiebe. "Seeing like a Peasant: The Origins of Food Sovereignty." In *Food Sovereignty: Reconnecting Food, Nature and Community*, edited by Hannah Kay Wittman, Annette Aurelie Desmarais, and Nettie Wiebe, 1–14. Halifax: Fernwood, 2010.

Operationalising food sovereignty through an investment lens: how agro-ecology is putting 'big push theory' back on the table

Louis Thiemann

International Institute of Social Studies, The Hague, Netherlands

A central question in the current debate on food sovereignty concerns the concepts and approaches to assist and frame the operationalisation of its agendas for peasant-based agricultural development. Another is the search for inclusive methods and language to discuss these operational, 'territorial' agendas with potential constituents. This paper argues that both questions call for an investment lens, a complementary approach within food sovereignty that proposes and discusses investments rather than political demands. Decolonial epistemology will treat existing investment lenses critically; however, in doing so it also urges new perspectives on what constitutes investment, the categories of cost involved, and the measurements employed. In following the rationale of investment in agro-ecological theory and practice, the paper next argues that the reconstruction of 'big push theory' outside the 'modernisation' paradigm that once produced it is possible, and that formulation and discussion of big push strategies could reclaim a space within critical agrarian studies. Big push theory offers a frame for the consistent critique of 'silver bullet' development projects through the study of negative feedback loops; and a frame for the study of positive feedback loops, which crucially underlie the proposals of food sovereignty movements for broad, integrated changes in agrarian systems.

Operationalising development theories: food sovereignty through an investment lens

Scepticism towards the food sovereignty (FS) frame has focused on the feasibility of translating its principles and ideals into political practice and coherently addressing the class positions towards and within the changes in agrarian structures it envisions.[1] In this way it questions the frame's analytical capacities to convincingly deal with issues such as consumer food prices, the internal differentiation of 'the peasantry' or the obvious and not-so-obvious arenas of competition between 'competing sovereignties'.[2] Similarly the lack of concreteness is seen as a hindrance in the construction of 'beyond-local', large-scale support structures that can effectively take over (although in a vastly different manner) those functions within food systems that are now played by corporations and states.[3] Indeed, texts such as the Nyéléni Declaration, as well as the strategic declarations of La Via Campesina (LVC), have been crafted largely in idealistic language, and debate on a number of thorny issues has only recently taken off.[4] Some proponents and supporters of the frame argue that its 'depiction of an alternative is not a depiction grounded in the messy compromises of the here and now but rather in a fully-fleshed-out depiction of another world, which, the global food sovereignty movement (FSM) argues, is possible to build, and now'.[5] In this context I propose the complementary use of an 'investment lens' to concretise what FS means in practice, thus allowing proponents and sceptics to systematically engage the potentials as well as the reality of FS and to fill the 'blank' spaces within the framework. By finding paths to concentrate on concrete, 'territorial'[6] investments in ways that circumvent quantification traps, such an approach can also help arrange additional evidence in favour of peasant-based agriculture, beginning with the powerful argument that investments by small farmers, assisted by other actors, are the *cheapest* way to reach development goals, thus beating advocates of large-scale industrial agriculture at their own efficiency-rhetoric game.

This resonates with recent FSM efforts to transcend the critique of capital's investment logic (large-scale land investment, economies of scale, value-chain integration etc) with a complementary debate on 'alternative agricultural investments'.[7] FSMs have engaged these debates in forums such as the Civil Society Mechanism (CSM) of the Committee on World Food Security (CFS), which was opened after FSMs successfully pressured the CFS not to ratify the World Bank's pro-business 'Principles for Responsible Investment in Agriculture'. Agro-ecology is seen as a major catalyst in this process, given its focus on collecting, developing and horizontally disseminating agricultural techniques based on the principles of low-input use, resilience, sustainability, as well as its prioritisation of smallholders. Recognising the original, more technical definition of agro-ecology as a scientific and/or farmer-led process of applying ecological science and knowledge to the study, design and management of agricultural systems,[8] FS has developed a more encompassing and political understanding of agro-ecology's purpose. In the context of multiple appropriations of agro-ecological science and techniques by competing actors, and for different purposes (following the recognisable patterns of earlier frames such as 'organic farming'), FS proponents have explicitly adopted agro-ecology as a tool for finding techniques that can enhance peasant autonomy within and against the hegemony of capitalist relations in agriculture.

Translating ideals into practice means advancing from visions of desirable states to the design of manageable projects for investment. Although recent history has loaded the term 'investment' with bitter references and irritable connotations, there is a powerful re-conceptualisation emerging. Peasant movements are pushing for broader acknowledgement of the fact that the vast majority of investment in agriculture is planned, carried out and paid for by farmers themselves, not by those commonly referred to as 'investors' or the state. FS contains its own investment paradigm – in fact it would be rather toothless without one, but has not done enough to spell this out. Taking an investment approach to FS would mean investigating the 'How to' question, a question that cannot be answered *universally* beyond vague, idealistic terms, but rather seeks the translation of its theories into a particular landscape, and into practical terms of change. Employing an investment lens thus powers efforts to territorialise development theories and agendas.

As a paradigm that emphasises the heterogeneity of peasant production and thus eschews painting generalised pictures of 'the agro-ecological farm' or 'the sustainable regional food system', FS is conditioned to institutionalise a relative separation of two layers: (1) the layer of general notions and abstract analysis of food systems, where talk is about directions and principles; and (2) a second layer on which territorial proposals are advanced, experimented with and evaluated, and where talk is about concrete investments. Translation from the first to the second layer (territorialising/'grounding'), and re-information in the opposite direction, are crucial processes, and their structuring through clear terms is key. The first translation process (operationalisation), in particular, however, cannot be left to the intuition of the respective translator only. Operationalising a vision does not mean going into a locality and starting to do something that intuitively seems to resonate with that vision. Rather it means finding theoretical and analytical tools and automations that can assist, but also take the weight off of that intuition. Consequently it means translating FS's objectives from political into practical language.

This paper discusses two potential tools to guide operationalisation: the first part outlines how FS's investment lens can be based on peasant farmers' own methodologies, thus resonating with FS's critique of the economistic valuation of agrarian development projects. The second part, in turn, shows how the theory of negative and positive feedback loops (applied in the development context as 'big push theory') can highlight some of the strongest points in FS's investment agenda. While FS itself could be considered a theory of *why* to invest in peasant-based agriculture, and agro-ecology (among other FS practices) a process of defining *what* exactly to invest in, the paper thus makes two points on *how* to frame investment into FS.

The humbling of science for agricultural change: the context for reconstructing the investment lens[9]

A central argument made by FS proponents is that the investment and input costs of agrarian development via peasant-based, agroecological and social justice strategies are markedly lower than those encountered by competing (green revolution) strategies, allowing them to be realised more frequently on existing small and medium-sized farms.[10] This argument is offered especially where the significant sums of external (private or state) investment necessary for green revolution strategies cannot be collected, or would come with strings attached. Many political-economy-related arguments for FS (such as the emphasis on peasant cooperatives and farmer-to-farmer networks as the main actors of change) also boil down to a reduction of investment costs as opposed to conventional strategies, as well as a change in their composition away from money and towards work and knowledge investment, thus making possible the envisioned substantial shift to capital-poor agents.

Behind this stand a number of epistemological changes that have 'humbled' the contribution of positivist science to agricultural change. These changes are the result, in part, of greater recognition of the limits of 'scientific' development strategies, in part of the evolution or recovery of cheaper, more sustainable or more democratic alternatives. FS breaks systematically with the assumption that policy makers, scientists and capitalists are the main holders of agency for agrarian change. Previously farmer families, small processors, artisans and other rural people were conceptualised as 'those affected' by engineered changes in incentive structures, a role they maintained within a large variation of paternalisms in both capitalist and socialist academies. But while the humbling of markets has been the paradigm of many successive Marxisms, and while the humbling of government has had a long tradition in cooperativist and radical libertarian thought, science and technology for agricultural change have only recently been humbled consistently, namely in agro-ecological theory and practice. This humbling process took place largely thanks to the dominant contribution of farmers' experimentation in the foundation and scaling-out of agro-ecology, which ignored and eventually dissolved previous hierarchies of knowledge production. If anything, many knowledge-chains saw their *a priori* hierarchies reversed towards assumptions and ideals that viewed farmers as the main agents of change, with academics holding a supportive function. While radical academics continue to be needed as 'defenders' within colonial knowledge structures as long as these exist (as critics and advocates of alternative knowledges), this reconfiguration of our condition has also made it possible to again develop a strong proactive focus in our work, ie to more openly become 'practitioners by other means'.

The changes in preconceptions of the 'ideal' plant breeding mechanisms as proposed by radicals may serve as an instructive example. Until the 1980s even many peasant advocates proposed centralised, 'scientific' breeding, although they wanted it under public ownership and internationalist exchange. Since then many of the hierarchies involved have been flatly reversed. The Programa de Inovación Agropecuaria Local (PIAL) in Cuba exhibits some of the new assumption about what is an ideal breeding mechanism. A project that by now stretches over 15 of Cuba's 16 provinces, it is partly responsible for the steady increase in seed diversity in the country since the 1990s. PIAL was built on the recognition that centralised breeding in national institutes would discourage seed diversity in the field, as it had in Cuba's period of agricultural modernisation (1960s to 1980s).[11] In the context of a national push to scale agro-ecology out into diverse local agro-ecosystems this was perceived as an inappropriate option. Instead, PIAL has found ways to organise the judgment of seed quality de-centrally, an act performed by farmers within their local agro-ecosystems, not by geneticists in laboratories.[12] Their chain of investments has proven cheap, flexible and effective. Through collection missions, *criollo* (native) varieties are collected throughout the country, and in combination with commercial seeds from the country's existing seed banks they form the stock of the project. Plant variety fairs are then organised at a local farmer's plot, each exhibiting between 40 and 150 varieties of a crop. Farming couples select a number of varieties with which they continue to experiment on their farms. Through this process germ plasm is not only scaled out into diverse agro-ecosystems with diverging requirements that only farmers can judge, it is also cross-bred and augmented by some of those farmers, leading to new varieties adapted to specific agro-eco-systems. The resulting biodiversity is stored in the extensive seed libraries main-tained by some farmers, and exchanged via farmer networks within or connected to the national small farmers union (ANAP).[13]

PIAL's seed system follows a number of the new epistemological ideals for research in the agro-ecology paradigm. These insist that the major part of research questions must be attacked in the field, not in the laboratory, and that they must be treated anew in each agro-ecosystem, given that agro-ecological solutions are not easily transferable from one situation to another. Since these requirements overburden the capacities of 'scientific' researchers, including their knowledge of diverse field conditions, most questions are only sensibly attacked through what has so far been called 'auto-innovation', ie farmers experimenting on their own, exchanging on their own or with a little 'academic' assistance.[14] Arguably the farmer-to-farmer network championed by some of LVC's member organisations is the only format capable of encouraging, facilitating and dissemi-nating endemic research on a larger scale, while internalising the pride of setting out to do research as well as promoting the findings.[15]

Justified objections to a quantitative investment lens

The idea of finding an 'investment lens' for studies in FS is likely to sound like a dangerous intrusion into the field, committing to a quantitative methodology that tends to produce economistic, money-focused perspectives on agro-ecosys-tems that are completely inadequate at capturing complex realities. In fact, one of the main vocations of FS has been to defend peasants against a set of propos-als informed by quantified 'investment costs' as the main focus, whether they hit rural communities in the form of Reducing Emissions from Deforestation and Degradation (REDD) schemes, or via the quantification of land values in land markets. Two negative tendencies go hand in hand when studies, proposals or policies are driven by a quantitative focus on investment and its costs: the omission of items that cannot be adequately quantified, and the many errors if quantification is attempted anyway (exemplified by the attempts to quantify poverty or hunger).

A way out has always existed but dominant epistemology before the 'decolo-nial turn' did not permit its use in 'scientific' research.[16] That way out is to re-frame investment costs as a wider category. This category would include three kinds of costs: (1) costs quantifiable in money(s); (2) costs only sensibly quanti-fiable in other terms (hours of certain types of work, units of a specific material, kWh of electricity); and (3) costs that are qualitative in nature (and must remain so), for example a political concession by a specific actor, security of land ten-ure, 'challenges' as they are perceived by a specific agent, or the skilling of workers in a specific subject. In a hierarchical academy the perceived flaw of this approach was that two proposals that used it could not be compared by a 'neutral', 'scientific' outsider – the academic. To achieve comparability, all variables needed to be connected to one or two root variables, usually money on the 'educt side' and yield (or again money) on the 'product side'. Even though complex statistical programmes have with time allowed more variables to be added on both sides of the equation, they needed to stand in a mathematical relation to each other in order to allow comparisons. Only then, it was argued, would the assembled evidence have 'comparative power'. In this tight frame an inclusive conceptualisation of 'investment' remained impossible to operationalise.

After the decolonial turn it is precisely this 'flaw' that becomes the greatest strength of an inclusive approach: since proposals for change are now, ideally, to be developed, evaluated and compared endemically (ie by their subjects), their comparative power, to use the same term, derives primarily from the facility and lucidity with which farmers and other practitioners can use the contained information in their endemic processes of evaluation and juxtaposition, understanding them in their own languages, evaluating them by their own standards, and weighing the variables in their own location and circumstances.[17]

Distilling and expressing investment costs while avoiding quantification traps: learning from farmers' methodologies

The aim of an investment proposal in agro-ecology is thus to assemble a display of information relevant to the agents addressed. I propose that simply listing the investments that form part of the proposal is an adequate format. It displays the given information without precluding the process of evaluation and judgment, recognising that the outcomes of the study then become the inputs to a second study that takes place outside the reach of the academic. Investment costs are expressed in their 'home' variables without venturing into troublesome conversions, and listed as a set of requirements to materialise a given development strategy in a given agro-ecosystem—some in terms of money, others in terms of hours or resource units, again others in qualitative terms. If comparative power is lost because the study can no longer speak for itself, it reappears at the opposite end, where the study begins to speak to those it wishes to persuade. Studies conducted during a number of PIAL plant variety fairs indicate that farmers proceeded in a similar way when gathering information on a proposed investment. When evaluating and selecting varieties, they saw the plant with a list of attributes, comprising both quantitative variables and qualitative impressions.[18] Farmers were 'reading' the exhibited varieties as a list of attributes/items, feeding the information into their own methodologies for evaluation (Table 1). And even though they used a large variety and variation of items, they were reported to select 'quickly and with ease'.[19]

The programme's role was merely to provide these lists without making any judgments in the process, which involved sowing the varieties in small plots in a homogeneous field. The academics involved consciously refrained from any form of 'ranking' or reprocessing of the information given.

Having an overview of necessary investments in hand, the political economy of each individual as well as combinations of proposed investments can be discussed, giving way to a dialectical spiral through which the initial proposal is

Table 1. Farmers' reading of varieties of maize at a seed fair in San José de las Lajas, Cuba.

Attributes	% of farmers using	Variable quantifiable?
Yield	87.5	Yes
Plant height	87.5	Yes
Leaf position	62.5	No
Number of leaves	60	Yes
Leaf colour	45.5	No
Leaf size	41.3	Yes
Stem width	76.3	Partly
Number of ears	57.5	Yes
Colour of ears	32.5	No
Size of ears	40	Yes
Weight of ears	50	Yes
Signs of fusariosis ('encamado')	31.3	Partly
Height of ears	40	Yes
Roundness of ears	40	No
Colour of pods	28.7	No
Diameter of ears	37.5	Yes
Cover of ears	55	Yes
Form of ears	55	No
Insect damage	35	Partly
Length of ears	45	Yes

Source: Data adapted from Rios, "La Diseminación Participativa de Semillas."

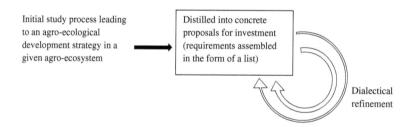

Figure 1. General path of an investment proposal.

further specified and reconciled with its circumstances (Figure 1). Via the same process it is adapted to changing political-economic circumstances.

For in-depth case studies the list format accommodates not only a first column of proposed investments, but also a second column that displays possible means by which these individual investments can be undertaken. The second column collects the costs associated with an investment, be it in money, hours, or a concession by a political actor, and represents a second layer on which arguments for agro-ecology can be made. In situations where the same investment can be undertaken in different ways, this second column grows in importance. For example: a function needed in a particular proposal (eg mechanical compost turning) can either be brought about by importing a machine from an international corporation, resulting in costs of $10,000 to the buyer, plus dependency for spare parts and maintenance, plus the costs to the domestic economy of generating $10,000 of foreign currency earnings; or it can be solved by building a similar machine from open-source blueprints and through national or local manufacturers, resulting in costs of $400 for imported parts, an amount of national currency during the manufacturing process, skilling of X workers, and a final cost to the buyer. In both cases the compost rows are turned, ie the investment is accomplished; however, the costs and the actors that bear them differ.

This distinction is enormously important when trying to show the potential of including alternative knowledge in neighbouring disciplines in a proposal. In agrarian change the topics of alternative ('appropriate') technologies and alternative (vernacular) architecture in, particular, may prosper within this format, as the reduction and restructuration of costs (both economic and social) they provide is much more visible and persuasive. In an investment proposal these secondary variables become integrated means to reach a goal, instead of being 'additional goals'.

One last aspect has to be considered: when thinking of investments not only additions to but also subtractions from the status quo have to be considered; indeed, sometimes they may be crucial items on the list. The reduction of the burden of weeding, for example, is a major argument for an agro-ecological transition in many agro-ecosystems (with due exceptions in the transition from high-pesticide systems). On the other hand, the loss of independence (in its respective local expressions) and/or of soil fertility must be entered as items on the list of many competing proposals, often including the business-as-usual proposal. This also holds in cases where only part of an investment cost is quantifiable, as is the case with all annual input costs whose effects are never adequately expressed in inflation-adjusted money terms but include dependences and loss of subsistence.

The investment lens and the creation of interfaces between academics and practitioners

For any individual creating as well as understanding abstract treatments of agro-ecosystems (including their social and political components), whether radical or 'mainstream', is a challenge. It revolves around particular cognitive abilities within what the theory of multiple intelligences has termed the logical and spatial intelligences.[20] Probably more importantly it requires a particular kind of patience – waiting for solutions to follow a long process of study in which the person spends time 'in the abstract' (ie learning and handling concepts that do not reveal their practical significance until much later, such as the different parts of a theoretical framework). Abstract treatment often represents a lengthy detour of a kind that fascinates some, but discourages most. Hence it must also be understood as a process of selection, which bolsters the position of those who have time, resources and the will to dig into systems of concepts, while demobilising all others. This has political implications for power differentiation, hierarchies and pride which are the concern of a growing body of decolonial literature. However, the main point here is much simpler: when a movement takes significant detours of abstraction to get from a reality to an envisioned future, it can lose large parts of its possible constituency to a mixture of boredom, passivity and impatience. If we assert that there are different ways of knowing, there are also different ways of imagining elements of a possible future. Although some movement members are most comfortable with imagining these in abstract terms, the majority of food sovereignty's potential constituency will prefer concrete forms of imagination, as well as concrete language to express problems and solutions. They will want to draw agro-ecology into the landscape they know, the farm they are responsible for, and the social fabric they are embedded in, in compliance with their experience, and in using their respective intelligence.

An ongoing debate in the field concerns the successful rapprochement of academic and practitioners' ways of proposing changes to an agro-ecosystem. If academics use an investment lens to distil their proposals, they build an inter-face on which practitioners and academics connect their respective research agendas. Around such tables a 'culture of proposing' can develop, with different parties seated at different angles (perspectives) to the proposal. I argue that the format of the investment proposal encourages more people to bring in their methodologies and talents, since:

(1) Agro-ecosystem changes are proposed in common language(s) that enable more people to take part in the culture, if they wish to do so. This 'down-to-earth' nature creates inclusive knowledge and facilitates participatory and practitioner-led research.
(2) Seeing changes specified in straightforward, palpable items also draws 'realist' and sceptically minded people into the culture.
(3) The investment focus pushes proposals towards depth (by going into the 'second column' of the list).
(4) An open methodology allows a larger variety of contributions to the debate (drawings or videos, but also pioneer plots and farms as live exhibitions of a proposal), as well as ways of leading it (not only during times and in spaces reserved for discussion, but also throughout the everyday life, work and contemplation of the agro-ecosystem, during casual encounters, in families and bars). Participatory Action Research (PAR) methods would gain in utility for FSMs and their allies, given that investment-based territorial imagination and proposals could be recognised more strongly within movements as well as within development academia.

All these aspects assist in building a culture of proposing, as exhibited tentatively by the first networks that have pursued their implementation in their organising structures, such as the Campesino a Campesino agro-ecology network(s) in Central America and Cuba.[21]

Finding the right scope for investment – a big push theory for food sovereignty?

Since the early critiques of the so-called 'green revolution' and conventional agricultural development (AD) in the 1970s, the development of an alternative agricultural investment paradigm has come a long way. In the coming decades the multiple crises of conventional agro-ecosystems will attach renewed immediacy to the question of their change, and new critical junctures for institutional change can be expected,[22] with different 'solutions' bidding for funds and recognition.[23] In this context agro-ecology is positioning itself with the call for a 'new agrarian revolution' through ecological agriculture and social justice.[24] But while its strategies at the farm- and village-scale are already well-defined and presentable, a theory and term (catchword) to describe the required effort at a national level has not yet emerged. This hinders the accumulation of evidence on the synergies between particular investments on different scales, levels and by different actors. In this context I argue for the need to reinvent, within the FS paradigm, the mental image of the 'big push', the concerted effort to break through negative feedback loops that was so present in early postcolonial development strategies. Big push theory is a particular approach to *how* investment into peasant-based agrarian development should be designed, which, I argue, can connect with the *why* given by FS theory and the *what* given by agro-ecology. Such a mentality is not in itself a question of scale (it is applicable to systemic changes from the farm to the national scale); rather it is about the scope of the solutions we propose. To change any non-lethargic system, a systemic approach informed by big push theory and operationalising big push practice is appropriate. Its antithesis is the silver bullet approach that focuses on single-intervention, often short-term solutions and remains relatively careless about feedback loops. Nonetheless, in the current FS debate, its first and foremost application is certainly at the public-policy level (national and regional scales), as evidenced below in the example of Brazil's family farming laws.

Negative feedback loops

The key insight of big push theory is the observation that economies with low capital stocks (physical, social, human and/or environmental capital) are subject to an array of feedback loops between their various 'dimensions of poverty'. This holds true regardless of their scale, from the farm economy to the national economy. The interplay of these feedback loops keeps them in a state of quasi-equilibrium at a low level of productivity, forcing in turn a low rate of savings and investment. In economic terms, these 'poverty traps' are characterised as inhibitions to increase the rate of investment out of the total of economic activity, be it to build the fertility of soils and use time and energies to experiment and learn, at the farm level; or to construct factories for equipment, lay out infrastructure or fund veterinary faculties, at the regional or national level.[25]

This basic logic also underlies conceptualisations of agrarian change through agro-ecology. In agro-ecology the conceptualisation of the agro-ecosystem as including not only local 'fields and farmers' but also regional to international policies, markets and climates, contains this perspective, and agro-ecologists' research has often followed a political economy/ecology approach to the study of land management. Although the term 'feedback loops' itself is not commonly used, every time a researcher finds that a problem manifests itself locally and is enacted by local actors, but decides that a purely local solution is unlikely to exist (ie the problem has roots in regional, national or international policies, markets, social systems or climates), s/he is uncovering negative feedback loops

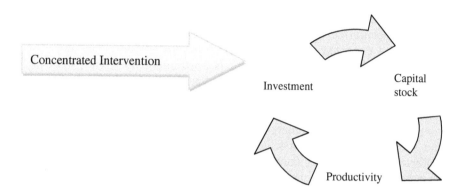

Figure 2. The central goal of big push strategies is an artificial increase in the investment rate.

that hinder or limit local change in the absence of changes in super-local structures.[26] The concept of resilience describes the quality of physical, ecological, human and social capital within an agro-ecosystem in light of its ability to withstand shocks.[27] It calls for the formation of such capital in order to build local and national sovereignty against environmental degradation and catastrophe and against economic exploitation and social differentiation. 'Poverty' is thus defined in socioeconomic *and* environmental terms, ie the poor are the subjects of agro-ecosystems characterised by high volatility, low yields relative to local needs, and/or high levels of inter-human exploitation. In such agro-ecosystems ecological as well as social and economic feedback loops are likely to limit the potential gains of small-scale projects.

A development proposal employs big push theory when it (a) recognises the (partial) 'trap' character of a status quo, (b) argues that, in order to leave a 'trap', several negative feedback loops need to be addressed in conjuncture via the coordination of investments at different layers, and (c) proposes solutions that function as positive feedback loops by multiplying the effects of the actual intervention. Finally, it recognises the necessity to institute a new 'network of negative feedback' to make the new system adaptive, as well as resilient against capital attrition and environmental and social shocks.[28]

The first life of big push theory until the 1980s – what lessons to draw?

Unquestionably the 'big push' notion was in the past primarily an appendix to large-scale green revolution and/or state capitalist development strategies, and has lost credibility together with these.[29] However, critical scholars and policy makers should not read the previous attempts to materialise a big push as a failure in theory. Rather, it was the implementation of a correct imperative under incorrect parameter assumptions, such as the focus on inputs, centralised knowledge creation and dissemination, and the progressive integration of peasants into spheres of market or state influence.

Green revolution-inspired proponents of the big push were, of course, never entirely true to their theory, and often created new negative feedback loops while attempting to 'solve' existing ones. Certainly any researchers' imagination of feedback loops and solutions will distil from the particular combinations of concepts and ideals present in his mind. In other words, a big push can be made for any purpose and, when a particular purpose develops its own big push theory, it may gain force but not validity.

As political movements throughout the developing world were nurtured on the vision of swift economic growth and development, the notion of the big push left the economics departments and developed an extensive *Eigenleben* (or life of its own) as a development 'mentality'. Many of these movements, however, failed to yield the expected results, developed autocratic features or lost ideological and military struggles. As a result, varieties of development pessimism flourished. Since the 1990s mentioning the agrarian big push is typically seen as a symbol of good intentions, but ultimately naivety, unworldliness or nostalgia. This did not change significantly when, during the early 2000s, the big push was rediscovered by scholars like Jeffrey Sachs and Paul Collier, and implemented through the United Nations Millennium Project.[30] Agricultural development (AD) efforts within such fast-track projects remained focused on input distribution and external 'training' of local farmers,[31] and should largely be understood as reincarnations of the 'first life' of the big push.[32]

As the progressive re-conceptualisation of investment is underway, the notion of the big push deserves to be reconsidered in the FS context. I argue that, first, big push theory offers a frame for the consistent critique of 'silver bullet' development projects. A key element in this frame is the concept and study of 'negative feedback loops' that dilute, diverge from, or even reverse the benefits of such projects in many contemporary agrarian settings, for example by accelerating social differentiation or increasing input dependency. Second, big push theory offers a frame for the study of positive feedback loops, which must crucially underlie the proposals of FSM in the context of constrained funding for agrarian development.

Why, and how, agro-ecology brings the big push back to the table

Agro-ecology responds to the resource-intensity of previous agricultural development strategies by focusing on innovations that can provide significant benefits at low external cost, and promoting knowledge and social capital above material inputs.[33] Thereby it constitutes investments that can be made by peasant farmers with little monetary resources, and that can be scaled-up by resource-constrained states, NGOs and farmer networks.

The significance of the adoption of agro-ecology by FSMs is that the possibility of solving problems via agro-ecological, instead of technological, solutions constitutes a radical reconfiguration of the investment needs with which common benefits can be achieved. Agro-ecology makes FS 'cheaper' to get, by shifting importance towards low-cost, knowledge-based and farmer-manageable investments that curtail debts (monetary as well as environmental). It does so by seeing agriculture as interconnected systems on different scales, framed around the idea of the 'agro-ecosystem' that contains ecological, social and political spaces within it. The integrated application of such investments thus becomes possible even in the absence of large-scale public funding.

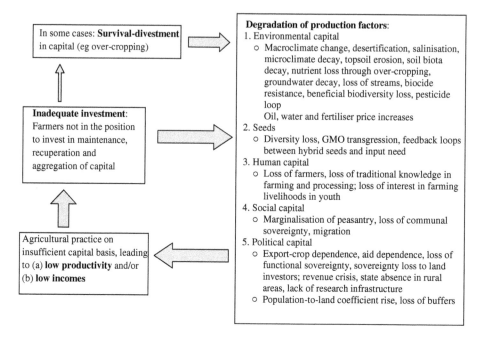

Figure 3. Nation-scaled low-sustainability feedback loop impeding the development of sustainable agriculture in developing countries in the absence of an agro-ecological big push.

Figure 3 shows FS's common diagnoses on failing agricultural systems in the form of a negative feedback loop that shows the restraint of the peasantry in adverse conditions. FS offers strategies on all five dimensions that keep the loop running. These include, among many others: (1) farming and conservation strategies that increase ecological capital and lower fossil fuel dependency; (2) open-source seed strategies, including diversification, farmer-led breeding and bans on genetically modified organisms; (3) strategies that secure the livelihoods of peasant farmers; (4) re-vesting communities and regions with sovereignty over their food systems; and (5) national food sovereignty strategies centred on a reactivation of the state.

When comparing 'conventional' and agro-ecological strategies, it becomes clear that they rely on different categories of investments, thereby generating different costs. To increase soil fertility, for example, agro-ecology aims at investments in the farm work flow (tillage reduction, mulching, livestock integration), introduction of new species (cover crops, green manure, fertiliser trees) and ecological techniques (succession, terracing, crop rotations), while conventional investment focuses on external fertilisation and herbicides. The former focuses on one-time costs (dissemination and adaptation of self-renewing knowledge and species, as well as labour investments to set up contour swales, nurseries, terraces etc), while the latter primarily features recurring, annual costs (inputs). This basic divergence has far-reaching effects on policy design, and it recurs in other dimensions of agricultural investment, such as labour productivity increases, market access, pest management and water management.

The Programa de Aquisição de Alimento (PAA) in Brazil: central programmes, local power

The recent revisions in Brazilian public procurement policies make an important point about the design of food sovereignty-based big push strategies, and how scale can be achieved without centralisation. One insight from Brazil's policy to shift public procurement programmes towards family farms (beginning in 2003) is that even programmes that are pushed and directly financed by large-scale institutions can be designed to have a localising, democratising effect. In 2012 the programme included 185,000 family farms who each sold an average of R $4554 (US$2058) worth of crops to 17,988 registered public and non-governmental agencies.[34] In practice procurement contracts are made between the local buyer and local farmer associations or individual farmers, while the central state agency overseeing the enormous programme (more than $450 million annually) is reduced to checking contracts and sending payments. This has resulted in operating costs of only 1.1% of total sales through the programme.[35] It has also created a dynamic through which potential actors in local food systems coordinate their respective needs and potentials, an ongoing exercise that has been shown to drive diversification of production and small-scale processing, as well as the recovery and strengthening of farmer organisations and local food cultures.[36]

One study showed that, even though the programme operates with regional market-referenced prices, replacing intermediaries with direct sales contracts increased farm gate prices by 45.9% on average.[37] Another used data from the official household survey to show that family-labour farms, whose main buyer was public or cooperative, had an average income of R$1361 in 2011, while those who sold to intermediaries earned R$493.[38] Because PAA contracts are fixed over a long term, 'the farmer is better placed to manage risk and is substantially protected from unfavorable fluctuations in market price and demand'.[39] Farms were found to increase their production specifically to sell to PAA.[40]

Many offshoots of PAA have begun operation in Brazil, both widening it in scope and adapting it to local conditions. Beginning in 2008, for example, the national agency overseeing the school feeding programme of 45 million students throughout the country implemented a 30% family farm quota for school procurement, a figure that was reached on a national scale in 2012, representing US $422 million of sales for more than 104,000 family farms.[41] Since 2012 a Food and Agriculture Organization–World Food Programme initiative with support from Brazil has been conducting pilots in five African countries to probe the possibility of adapt the PAA method in different contexts; the Brazilian Development Corporation has used the methodology to revitalise milk production in Haiti.[42]

Imagining the big push and the psychology of social movements

Thus far the argument has been that the big push did not leave us in the 1970s, but that it was waiting for a general overhaul while being parked by the forces corresponding to neoliberal globalisation. For multiple, interlinked reasons a new opening for the big push mentality may develop. Agro-ecology can, by devising its own big push strategies, take a stand in the debate on the parameters of that next wave – what kinds of big push are possible on a national scale and how they could be unfolded. This necessitates not only a critique of conventional strategies on the national scale, but also proactivity in proposing alternatives for the same scale.

A strong argument for proactive, ie pre-emptive, proposals is found in social movement studies. Whether potential participants of agrarian movements, unions, parties, community organisations and other change-oriented groups join or remain passive outside them; whether they invest time and effort or remain passive inside them; and whether they face or flee repression depends on a number of pull- and push forces. Of those forces that the movement itself can control (excluding therefore state repression and media stigmatisation), one of the most potent is the perceived *cause* of the movement. The 'institution' of the cause spurs members and sympathisers to imagine possibilities that the movement's success could make real. In the 20th century, for instance, the individual and collective exercise of imagining a communist society contributed enormously to mobilisation and sympathy for the respective movements, arguably far beyond the 'objective' class interests of those who joined.

For emerging agrarian movements and their supporters developing proposals for an agro-ecological big push in their territory could have a number of effects on the radiance of their shared cause. It can integrate existing partial causes (eg resistance to specific projects or processes, advocacy of specific laws), thereby discussing how such integration is feasible, and potentially bridging circumstance-induced conflicts between member factions. Likewise big push strategies proposed by or in collaboration with academics give credibility to the cause (by conducting feasibility calculations and extrapolating gains from pioneer cases onto the national scale). And, in doing so, they increase the perceived potential of pioneer projects or educational efforts, making them more valuable activities to engage in. Examples of these effects are the studies coordinated by Eric Holt-Giménez on the resilience of agro-ecological pilot farms in Central America,[43] and the work of Jules Pretty on the yield gains and input need decrease in agro-ecological pioneer projects across the globe.[44] Both gave an important argument to agrarian movements by convincingly underlining claims that peasants, using agro-ecological techniques, can feed the world.

For most current as well as potential members and sympathisers, a lower abstractness of long-term visions fosters not only their comprehension, but also the ability to discuss, critique and thereby scale these visions out into people's specific contexts. The collective educational process that is one pillar of movement building can thus be vitalised. By evolving the debate over a movement's cause from abstract concepts ('sustainable agro-ecosystems'; 'equality'; 'peasant rights', etc) to actual landscapes, its 'imaginability' increases, encouraging the debate over it to include not only intellectual and political circles, but virtually everyone who, by living in a particular landscape, has intimate knowledge and experience of it. The 'cause' becomes concrete: formable and debatable with techniques such as drawings on paper or sand, and using local terms, crops, poly-cultures, currencies and other calculations of value, extrapolated to local geographies.

Conclusions

Concretising FS through an investment lens helps to tease out its quintessence in situated terms and understandable language, and allows movements and their allies to territorialise the ideals of FS into practice by developing, discussing and scaling out investment proposals. There is a need to structure the evidence about what investment in food sovereignty and agro-ecology beyond the farm mean, with a focus on the synergies (or allergies) between efforts in different spaces and by different actors.

This is especially, though not exclusively, the case for questions of scaling up to the national level. Big push theory problematises the synergies between action on different levels and by different actors, and it can do so for systems of any scale. Although famous for justifying the use of chemical technologies in the 1960s and 1970s, in its internal logic it always contained an ecological theory of change, eschewing silver bullets and viewing change through an eco-systemic lens. It can be argued that the narrow objectives it received in its first life obstructed this characteristic. Agro-ecology is a much more suitable content-giver for big push methodologies than homogenising industrial capitalist agriculture was. The basic rationale of agro-ecological solutions is systemic – recognising and engaging ecological feedback loops. At the farm and local levels feedback loops are clearly starting points of FS practice, and the theory on farm-level agro-ecological transitions is about making 'the big push' on a small scale. Moore Lappé called this 'approaching change with an eco-mind'.[45] The same approach can guide FS's quest to develop its strategies for and at the national level and scale.

In addition to this, an argument can be made that focusing on territorialised visions of FS when promoting it beyond core constituencies can assist in safe-guarding concepts such as agro-ecology and FS against co-option. Lessons must be drawn from preceding conceptualisations of progressive agrarian change that left important questions of investment open or vague, and were susceptible to corporate capture as a result. The problem of concept-based agitation is that any concept can be misread or hijacked, whether from within or from outside a movement.[46] The alternative is to agitate for proposals that are specific, un-dilut-able and difficult to attack discursively – products of an investment lens.

Since the parameter assumptions that frustrated big push strategies in earlier incarnations have been exposed and replacements found and consolidated, the question of whether FS should, given the immediate necessity to transform agro-ecosystems on a large scale, formulate a revised framework for concerted action in the transition to sustainable agricultures needs to be addressed. Struc-turing the scattered knowledge about the feedback loops involved in unsustain-able agriculture, as well as the evidence from the many attempts to overcome them using agro-ecological and FS principles and techniques, means engaging in big push theory. Developing the big push dimension of FS means operation-alising how to 'change the system' beyond, and complementary to, the bigger-picture struggles, such as around land speculation and corporate value-chain integration.

Notes

1. Bernstein, "Food Sovereignty."
2. Ibid; and Schiavoni, "Competing Sovereignties."
3. Li, *Land's End*; and Burnett and Murphy, "What Place for International Trade?"
4. Taking place at the Food Sovereignty conferences in Yale, September 14–15, 2013 and The Hague, January 24, 2014. Issues such as class, the consumer–producer divide and international trade found great resonance at these meetings. See Bernstein, "Food Sovereignty"; Schiavoni, "Competing Sovereignties"; and Burnett and Murphy, "What Place for International Trade?"
5. Akram-Lodhi, "How to Build Food Sovereignty."
6. With 'territorial' proposals I refer to such proposals for investment that are specific to a place and circumstance, rather than 'blueprints' to be applied across the board.
7. Kay, *Policy Shift*; and Kay, *Positive Investment Alternatives*. See also HLPE, "Investing in Smallholder Agriculture."
8. Altieri, "Agroecology."
9. The term is taken from the formulation that decoloniality is built through the 'humbling of modernity' in Vazquez, "Towards a Decolonial Critique of Modernity."
10. Altieri and Toledo, "The Agroecological Revolution."
11. Ríos, "La Diseminación Participativa de Semillas."
12. Moreno, "El Fitomejoramiento"; and Ortíz et al., "Logros del Fitomejoramiento."
13. Ríos, "La Diseminación Participativa de Semillas."
14. Leitgeb et al., "Academic Discussion."
15. Holt-Giménez, *Campesino a Campesino*.
16. Escobar, *Encountering Development*.
17. Martínez-Torres and Rosset, "Dialogo de saberes."
18. Ríos, "La Diseminación Participativa de Semillas."
19. Ibid., 92.
20. Gardner, *Frames of Mind*.
21. Rosset et al., "The Campesino-to-Campesino Agroecology Movement"; and Machin Sosa et al., *El Movimiento de Campesino a Campesino*. See also Holt-Giménez, *Campesino a Campesino*.
22. Acemoglu and Robinson, *Why Nations Fail*.
23. Weis, "The Accelerating Biophysical Contradictions."
24. Altieri and Toledo, "The Agroecological Revolution in Latin America."
25. Rosenstein-Rodan, "Problems of Industrialisation."
26. The concept and study of negative feedback loops originates in species ecology, whence it was adapted to the study of agricultural systems and, finally, into political ecology. Clapman, *Natural Ecosystems*, shows that feedback loops are constituted by material flows of energy or nutrients in natural ecosystems. In human-controlled ecosystems this encompasses material flows of income and capital, as well as configurations of power, attention and incentives. While negative feedback loops diminish the effect of an intervention or shock ('resilience'), positive feedback loops amplify the effect, often in unintended directions.
27. Holling, "Resilience and Stability."
28. Ryszkowski, "Agriculture and Landscape Ecology," 342.
29. Big push theory first emerged in the late 1950s under the umbrella of the modernisation theory of development. See Ghosal, "The Theory of Big Push," 225; Rosenstein-Rodan, *Notes on the Theory of the 'Big Push'*; and Nelson, "A Theory of the Low-level Equilibrium Trap."
30. Sachs and McArthur, "The Millennium Project."
31. Sanchez et al., "The African Millennium Villages."
32. Wilson, "Model Villages."
33. Altieri, "Agroecology."
34. Sambuichi et al., "Compras Públicas Sustentáveis"; and CONAB, *A Evalução do Programa*.
35. CONAB, *A Evalução do Programa*.
36. Chmielewska and Souza, *Market Alternatives*; and Sambuichi et al., "Compras Públicas Sustentáveis".
37. Agapto et al., "Avaliação."
38. IPC, *Structured Demand*.
39. Chmielewska and Souza, *Market Alternatives*, 12.
40. Agapto et al., "Avaliação"; and IPC, *Structured Demand*.
41. IPC, *Structured Demand*.
42. Souza and Klug, "A Multidimensional Approach."
43. Holt-Giménez, "Measuring Farmers' Agroecological Resistance."
44. Pretty, "Agroecological Approaches."
45. Moore Lappé, "Beyond the Scarcity Scare."
46. Jaffee and Howard, "Corporate Cooption."

Bibliography

Acemoglu, D., and J. A. Robinson. *Why Nations Fail: The Origins of Power, Prosperity, and Poverty*. New York: Crown Business, 2012.

Agapto, J. P., R. S. Borsatto, V. F. de Souza Esquerdo, and S. M. P. P. Bergamasco. "Avaliação do Programa de Aquisição de Alimentos (PAA) em campina do Monte Alegre, estado de São Paulo, a partir da percepção dos agricultores." *Informações Econômicas* 42, no. 2 (2012): 13–21.

Akram-Lodhi, A. H. "How to Build Food Sovereignty." Paper for discussion at the international conference on 'Food Sovereignty: A Critical Dialogue', Yale University, September 14–15, 2013.

Altieri, M. A. "Agroecology: A New Research and Development Paradigm for World Agriculture." *Agriculture, Ecosystems and Environment* 27 (1989): 37–46.

Altieri, M. A., and V. M. Toledo. "The Agroecological Revolution in Latin America: Rescuing Nature, Ensuring Food Sovereignty and Empowering Peasants." *Journal of Peasant Studies* 38, no. 3 (2011): 587–612.

Bernstein, H. "Food Sovereignty: A Skeptical View." Paper for discussion at the international conference on 'Food Sovereignty: A Critical Dialogue', Yale University, September 14–15, 2013.

Burnett, K., and S. Murphy. "What Place for International Trade in Food Sovereignty?" Paper for discussion at the international conference on 'Food Sovereignty: A Critical Dialogue', Yale University, September 14–15, 2013.

Chmielewska, D. and D. Souza. *Market Alternatives for Smallholder Famers in Food Security Initiatives: Lessons from the Brazilian Food Acquisition Programme*. IPC-IG Working Paper 64. Brasilia: International Policy Centre for Inclusive Growth, 2010.

Clapham Jr., W. B. *Natural Ecosystems*. New York: Macmillan, 1983.

CONAB. *A Evolução do Programa de Aquisição de Alimentos*. Brasilia, 2013.

Escobar, A. *Encountering Development: The Making and Unmaking of the Third World*. Princeton, NJ: Princeton University Press, 2011.

Gardner, Howard. *Frames of Mind: The Theory of Multiple Intelligences*. New York: Basic Books, 1983.

Ghosal, S. N. "The Theory of Big Push in Indian Agriculture." *Indian Journal of Economics*, nos. 43–44 (1962): 225–232.

High Level Panel of Experts on Food Security and Nutrition (HLPE). "Investing in Smallholder Agriculture for Food Security." Committee on World Food Security, Rome, 2013.

Holling, C. S. "Resilience and Stability of Ecological Systems." *Annual Reviews on Ecological Systems* 4, no. 1 (1973): 1–23.

Holt-Giménez, E. *Campesino a Campesino: Voices from Latin America's Farmer-to-Farmer Movement for Sustainable Agriculture*. Oakland, CA: Institute for Food and Development Policy, 2006.

Holt-Giménez, E. "Measuring Farmers' Agroecological Resistance after Hurricane Mitch in Nicaragua: A Case Study in Participatory, Sustainable Land Management Impact Monitoring." *Agriculture, Ecosystems & Environment* 93, nos. 1–3 (2002): 87–105.

International Policy Centre for Inclusive Growth (IPC). *Structured Demand and Smallholder Farmers in Brazil: The Case of PAA and PNAE*. Brasilia: IPC, 2013.

Jaffee, D., and Howard, P. H.. "Corporate Cooptation of Organic and Fair Trade Standards." *Agriculture and Human Values* 27, no. 4 (2010): 387–399.

Kay, S. *Positive Investment Alternatives to Large-scale Land Acquisitions or Leases*. Amsterdam: Transnational Institute, 2012.

Kay, S. *Policy Shift: Investing in Agricultural Alternatives*. Amsterdam: Transnational Institute, 2014.

Leitgeb, F., E. Sanz, S. Kummer, R. Ninio, and C. R. Vogl. "Academic Discussion about Farmers' Experiments – A Synthesis." *Pastos y Forrajes* 31, no. 1 (2008): 3–24.

Li, T. *Land's End: Capitalist Relations on an Indigenous Frontier*. Durham, NC: Duke University Press, 2014.

Machín Sosa, B., A. M. Roque Jaime, D. R. Ávila Lozano, and P. M. Rosset. *El Movimiento de Campesino a Campesino de la ANAP en Cuba*. Havana: Asociación Nacional de Agricultores Pequeños, 2010.

Martínez-Torres, M. E., and Rosset, P. M.. "Diálogo de Saberes in La Vía Campesina: Food Sovereignty and Agroecology." *Journal of Peasant Studies ahead-of-print* (2014): 1–19.

Moore Lappé, F. "Beyond the Scarcity Scare: Reframing the Discourse of Hunger with an Eco-mind." *Journal of Peasant Studies* 40, no. 1 (2013): 219–238.

Moreno, I., V. Puldón, and H. Ríos. "El Fitomejoramiento y la Selección Participativa de Variedades de Arroz." *Cultivos Tropicales* 30, no. 2 (2009): 24–30.

Nelson, R. R. "A Theory of the Low-level Equilibrium Trap in Underdeveloped Economies." *American Economic Review* 46, no. 5 (1956): 894–908.

Ortiz, R., H. Ríos, M. Márquez, M. Ponce, V. Gil, M. Cancio, O. Chaveco, O. Rodríguez, A. Caballero, and C. Almekinders. "Logros del Fitomejoramiento Participativo Evaluado por los Productores Involucrados." *Cultivos Tropicales* 30, no. 2 (2009): 106–112.

Pretty, J. "Agroecological Approaches to Agricultural Development." Background Paper for the World Development Report 2008. RIMISP-Latin American Center for Rural Development, 2006.

Ríos, H. "La Diseminación Participativa de Semillas: Experiencias de Campo." *Cultivos Tropicales* 30, no. 2 (2009): 89–105.

FOOD SOVEREIGNTY

Rosenstein-Rodan, P. N. "Problems of Industrialisation of Eastern and South-Eastern Europe." *The Economic Journal* 53, no. 210/211 (1943): 202–211.

Rosenstein-Rodan, P. N. *Notes on the Theory of the 'Big Push'*. Cambridge, MA: Center for International Studies, Massachusetts Institute of Technology, 1957.

Rosset, P. M., B. Machín-Sosa, and A. M. Roque Jaime, and D. R. Ávila Lozano. "The Campesino-to-Campesino Agroecology Movement of ANAP in Cuba: Social Process Methodology in the Construction of Sustainable Peasant Agriculture and Food Sovereignty." *Journal of Peasant Studies* 38, no. 1 (2011): 161–191.

Ryszkowski, L. "Agriculture and Landscape Ecology." In *Landscape Ecology in Agroecosystems Management*, edited by L. Ryszkowski, 341–348. Boca Raton, FL: CRC Press, 2002.

Sachs, J. D., and McArthur, J. W.. "The Millennium Project: A Plan for Meeting the Millennium Development Goals." *Lancet* 365, no. 9456 (2005): 347–353.

Sambuichi, R. H. R, E. P. Galindo, M. A. C. de Oliveira, and A. M. de Magalhães. "Compras Públicas Sustentáveis e Agricultura Familiar: A Experiência do Programa de Aquisição de Alimentos (PAA) e do Programa Nacional de Alimentação Escolar (PNAE)." Mimeo, Brasilia, 2013.

Sanchez, P., C. Palm, J. Sachs, G. Denning, R. Flor, R. Harawa, B. Jama, et al. "The African Millennium Villages." *Proceedings of the National Academy of Sciences of the United States of America* 104, no. 43 (2007): 16775–16780.

Schiavoni, C. "Competing Sovereignties in the Political Construction of Food Sovereignty." Paper for discussion at the international conference on 'Food Sovereignty: A Critical Dialogue', Yale University, September 14–15, 2013.

Souza, Darana, and Israel Klug. "A Multidimensional Approach to Food Security: PAA Africa." *Poverty in Focus* 24. Brasilia: UNDP International Policy Centre for Inclusive Growth, 2012.

Vázquez, R. "Towards a Decolonial Critique of Modernity: Buen Vivir, Relationality and the Task of Listening." In *Denktraditionen im Dialog: Studien zur Befreiung und Interkulturalität*, edited by R Fornet Betancourt, 241–252. Aachen: Wissenschaftsverlag Mainz, 2012.

Weis, T. "The Accelerating Biophysical Contradictions of Industrial Capitalist Agriculture." *Journal of Agrarian Change* 10, no. 3 (2010): 315–341.

Wilson, J. "Model Villages in the Neoliberal Era: The Millennium Development Goals and the Colonization of Everyday Life." *Journal of Peasant Studies* 41, no. 1 (2014): 107–125.

Accelerating towards food sovereignty

A. Haroon Akram-Lodhi

Department of International Development Studies, Trent University, Peterborough, Canada

Rural social movements and urban food activists have sought to build food sovereignty because it has the potential to be the foundation of an alternative food system, transcending the deep-seated social, economic and ecological contradictions of the global food economy. However, continuing to build food sovereignty requires changes to global and local food systems that have to be undertaken in the messy reality of the present. This article therefore presents a series of wide-ranging, politically challenging but ultimately feasible interventions that are necessary but not sufficient conditions for its realisation.

The rise of food sovereignty

Food sovereignty has come to occupy a central place in the discourse of food activists. For an idea that emerged from a series of discussions among farmers who were members of Via Campesina, the global peasant movement, food sovereignty has mushroomed. As of October 2014, googling the term generated over 809,000 hits, a search on Google Scholar generated over 11,100 hits, and multilateral rural development agencies such as the Food and Agriculture Organization (FAO), the World Food Programme (WFP) and the International Fund for Agricultural Development (IFAD) employ the term in their discussions. Bolivia, Ecuador, Mali, Nepal, Senegal and Venezuela have embedded food sovereignty within their constitutions, and a diverse set of food-based civil society organisations have enshrined food sovereignty as a guiding principle. Food sovereignty has become part of the basic discourse of social justice advocates and organisations, including many that are not organised around food issues.

The rise of food sovereignty reflects a series of failures in the corporate food regime that has emerged over the past quarter century.[1] Dominated by global agro-food transnational corporations (TNCs), driven by financial market imperatives of short-run profitability, and characterised by the relentless food commodification processes that underpin 'supermarketisation', the corporate food regime forges global animal protein commodity chains while at the same time spreading transgenic organisms, which together broaden and deepen what Tony Weis calls the temperate 'industrial grain–oilseed–livestock' agro-food complex.[2] At the point of agricultural production the dominant producer model of the corporate food regime is the fossil-fuel-driven, large-scale, capital-intensive industrial agriculture mega-farm, which is in turn predicated upon deepening the simple reproduction squeeze facing petty commodity producers around the world.[3] A core market for agro-food TNCs are relatively affluent global consumers in the North and South, whose food preferences have been shifted towards 'healthier', 'organic' and 'green' products with large profit margins. At the same time, however, for the global middle class the corporate food regime sustains the mass production of very durable, highly processed food manufactures that are heavily reliant on soya, sodium and high fructose corn syrup, and whose lower profit margins mean that significantly higher volumes of product must be shifted.[4] Thus, the corporate food regime simultaneously fosters the ongoing diffusion of industrial agriculture – Fordist food such as McDonalds – as well as standardised differentiation – post-Fordist food such as sushi. The corporate food regime is sustained by capitalist states, the international financial and development organisations that govern the global economy, and big philanthropy. Missing from the profit-driven logic of the corporate food regime are those who lack the money needed to access commodified food and who are thus bypassed by the regime – a relative surplus population which is denied entitlement to food as a result of the normal and routine working of the global food system and who are thus subject to food-based social exclusion.[5] At the same time the corporate food regime is predicated upon a model of production, distribution and consumption that significantly exacerbates climate change and degrades the ecological foundations of the production upon which it depends.[6]

Although it was first developed to challenge the neoliberal globalisation being promoted by the World Trade Organization (WTO), the influence of the concept of food sovereignty has grown because it offers a different way of thinking about how the world food system can be organised; it offers an alternative. As developed initially by Via Campesina and further elaborated at the 2007 Nyéléni Forum for Food Sovereignty,[7] food sovereignty is based on the right of peoples and countries to define their own agricultural and food policy and has five interlinked and inseparable components:[8]

(1) A focus on food for people: food sovereignty puts the right to sufficient, healthy and culturally appropriate food for all individuals, peoples and communities at the centre of food, agriculture, livestock and fisheries policies, and rejects the proposition that food is just another commodity.
(2) The valuing of food providers: food sovereignty values and supports the contributions, and respects the rights, of women and men who grow, harvest and process food and rejects those policies, actions and programmes that undervalue them and threaten their livelihoods.
(3) Localisation of food systems: food sovereignty puts food providers and food consumers at the centre of decision making on food issues; protects providers from the dumping of food in local markets; protects consumers from poor quality and unhealthy food, including food tainted with

transgenic organisms; and rejects governance structures that depend on inequitable international trade and give power to corporations. It places control over territory, land, grazing, water, seeds, livestock and fish populations in the hands of local food providers and respects their rights to use and share them in socially and environmentally sustainable ways; it promotes positive interaction between food providers in different territories and from different sectors, which helps resolve conflicts; and rejects the privatisation of natural resources through laws, commercial contracts and intellectual property rights regimes.

(4) The building of knowledge and skills: food sovereignty builds on the skills and local knowledge of food providers and their local organisations that conserve, develop and manage localised food production and harvesting systems, developing appropriate research systems to support this, and rejects technologies that undermine these.

(5) Working with nature: food sovereignty uses the contributions of nature in diverse, low external-input agroecological production and harvesting methods that maximise the contribution of ecosystems and improve resilience. It rejects methods that harm ecosystem functions, and which depend on energy-intensive monocultures and livestock factories and other industrialised production methods.

Food sovereignty is thus an idea that is an alternative to 'food security', as this term has come to be used by multilateral development institutions to promote market-based and monetary solutions to a lack of access to food, and which says nothing about the inequitable structures and policies that have destroyed rural livelihoods and the environment and thus produced food insecurity. Food sovereignty instead offers a practice that is an alternative to the corporate food regime, with its proponents arguing that the food system needs to be predicated upon a decentralised agriculture, where production, processing, distribution and consumption are controlled by communities. So food sovereignty offers a vision of a real utopia rooted in the contemporary praxis of individuals and movements, and this vision has, in less than 20 years, become a critical component of global food movements.

Holt-Giménez and Shattuck have defined global food movements as the 'tens of thousands of local, national and international social movements concerned with food and agriculture', which 'have developed a wealth of political, technical, organizational and entrepreneurial skills, and advance(d) a wide range of demands',[9] as a consequence of which global food movements currently takes on a variety of guises.[10] Some are explicitly transformational, in the sense that they challenge the exploitative market structures of contemporary capitalism and the poverty produced by such structures. It is here that core support for the practice of food sovereignty can be found: Via Campesina, the International Planning Committee on Food Sovereignty, the Foodfirst International Action Network, and many food justice and food rights-based movements.[11] Some within global food movements are progressive, in the sense that they work within but are critical of the market structures of contemporary capitalism and the food-based inequalities that it creates: some fair trade chapters, some community food-security movement chapters, some community shared agriculture chapters, many slow food chapters, and many food policy councils. Within this group, many purport to support food sovereignty, without subscribing to the structural ramifications and necessary political practice of food sovereignty; for these progressives, food sovereignty is a rhetorical device rather than a political programme. Finally, there are those in the global food movements who promote more reformist visions of food and 'development': individual staff at international development institutions like the World Bank, the FAO, the IFAD, the WFP, the United Nations Development Programme, UN Women and the UN Commission on Sustainable Development; some fair trade chapters; some slow food chapters; some food policy councils; and most food banks and food aid programmes. At times these people may use the phrase 'food sovereignty' but when they do they have little understanding of the politics and practice of food sovereignty. Thus, it is important to emphasise that global food movements must not be equated with transformational movements for food sovereignty. Indeed, as will be discussed below, one of the challenges facing proponents of food sovereignty is to locate the interstices within global food movements where the practice of food sovereignty could be more deeply insinuated.

What does food sovereignty entail?

Food sovereignty offers an appealing alternative, and this helps explain its resonance. However, accelerating the processes and practices needed to achieve food sovereignty requires real changes to global and local food systems that have to be undertaken in the chaotic reality of the present. In what follows therefore a series of politically challenging but ultimately feasible interventions that are necessary but not sufficient conditions for the realisation of food sovereignty are presented. This is done because, despite its depiction of a real utopia, food sovereignty is grounded in the messy compromises of the practical political praxis of the here and now, which is a question about where social power is located and how it can be transformed. Transformational food sovereignty movements answer this question by addressing the production and consumption relations found in contemporary agriculture, because at the core of food sovereignty are new forms of social power – political, productive and market – which are being constructed around the production and consumption of food.[12]

However, the question of social power must be placed in the context of our times. The prevailing set of social-property relations within which food providers and food consumers are embedded is capitalist – the means of production are under the control of a socially dominant class, labour is 'free' from significant shares of the means of production and free to sell its capacity to work, and the purpose of commodity production is the seeking of profit. The localised smallholder farming model that is central to food sovereignty's alternative food system, and which, by continuing to be widespread throughout Asia, Africa and Latin America, might be thought of as an 'incubator' of food sovereignty, cannot be abstracted from capitalist social relations, which are defined by relations of exploitation between capital and classes of labour.[13] Smallholder farming is currently subordinated, through a range of mechanisms under the corporate food regime, to capitalist social property relations.[14] Thus, and as its transformational proponents clearly recognise, food sovereignty is not about trying to reconfigure the existing social conditions and relations of capitalism; it requires transcending the social conditions and relations of capitalism and developing a post-capitalist agrarian – and non-agrarian – alternative. Thus food sovereignty is about building power within the fissures of capitalist social-property relations, in order to transform food systems in favour of peasants, smallholders, fishers, food system workers and underserved communities. That this is the case is not recognised by progressive and reformist strands within global food movements, despite their use of the phrase 'food sovereignty'.

Food sovereignty is thus an objective which must continue to be sought, by popular mobilisation and struggle, by practice and by dialogic popular learning.[15] In this light it is worth asking: what is the path of change mapped out by transformational food sovereignty movements? Leaving to one side, for now, the role of praxis and teaching, there are two avenues deployed by transformational food sovereignty movements in their struggles:

(1) Mobilising against the policies and institutions hostile to the interests of peasants, farmers and workers, in order to propel change. To this end transformational food sovereignty organisations wage ongoing campaigns against land grabbing, transgenic organisms, agrofuels, the violation of the human rights of peasants, international food trade and aid governance, and the poor living standards of rural workers, and in support of agrarian reform, gender justice, improved terms and conditions of rural employment, action to mitigate climate change, indigenous rights, indigenous knowledge, improved rural nutrition, seed sharing and conservation, fisherfolk rights, and the UN Human Rights Commission draft declaration on the human rights of peasants.

(2) Negotiating with state institutions and international development organisations when it is believed that possible policy changes might be brought about through such collaboration. It was through this kind of process, for example, that food sovereignty was noted in the Final Declaration of the International Conference on Agrarian Reform and Rural Development in 2006,[16] following which the phrase starting turning up in FAO documents. The clear overlap in the interests and objectives of transformational food sovereignty movements and the Office of the UN Special Rapporteur on the Right to Food has also resulted in the phrase becoming more common in UN documents, particularly around indigenous rights.[17] Finally, transformational food sovereignty movements have been strongly engaged with the heavily reformed Committee on World Food Security (CFS), which is a collaborative venture of the FAO, the IFAD, and the WFP but which has significant input from global civil society organisations, including transformational food sovereignty organisations.

However, I would argue that in these struggles and engagements there is only a partial elaboration of what kind of specific changes would bring about an acceleration of pathways toward food sovereignty. It might be argued that the inability to define a wider range of specific changes that they favour reduces the ability of transformational food sovereignty movements to reshape the perspective of the progressive and reformist strands of global food movements, because it reflects an inability to identify the possible pathways by which societies can move into a more just future.

Pathways to food sovereignty

Another food system is possible. Indeed, given the role of the corporate food regime in accelerating climate change, the construction of an alternative food system is an urgent necessity.[18] Moreover, given the changes that have been wrought by neoliberal globalisation in the past quarter century, significant changes to the food system can probably be built within a comparatively short period of time – say, a generation. The question, though, is *how* can movement towards food sovereignty be accelerated? In what follows, I elaborate eight pathways towards food sovereignty.

Agrarian reform

The starting point in constructing accelerated pathways to food sovereignty must be the explicit policy change that is espoused by transformational food sovereignty movements: pro-poor, gender-responsive redistributive agrarian reform.[19] A 'stylised fact' of development is that at a global level the distribution of land and other rural resources is either the result of the thievery that was imperialism or of the glaring inequalities produced by market imperatives as capitalism was introduced into the countryside of developing capitalist countries by imperialist powers.[20] Granted, this stylised fact does not hold in all places and all spaces; but as a general statement it holds true in 2015. Pro-poor, gender-responsive redistributive agrarian reform is defined as a redistribution of land and other rural resources from the resource-rich to the resource-poor, such that the resource-poor are net beneficiaries of the reform and the resource-rich are net losers from the reform.[21] Pro-poor, gender-responsive redistributive agrarian reform may involve a plurality of social-property relations, including private and collective forms of property. Pro-poor, gender-responsive redistributive agrarian reform directly addresses the historical injustices by which farmers have lost their access to land over the past 150 years,[22] fundamentally tempers some of the glaring inequalities generated by market imperatives,[23] meets a basic precondition of how the rurally marginalised can begin to improve their livelihoods and thus address global poverty,[24] and creates the basis by which human rights can be realised and social and economic conditions transformed in a pro-poor direction.[25] Indeed, it is for these reasons that transformational food sovereignty movements consistently propose that pro-poor, gender-responsive redistributive agrarian reform is an urgent necessity for petty commodity producers around the world.

Moreover, a glaring aspect of development is that the foundation of structural transformation in East Asia was pro-poor redistributive agrarian reform, which brought forth the incentives to maximise agricultural production among the very poorest strata of society, who did so in order to create the preconditions of a better life for their families.[26] Granted, pro-poor redistributive agrarian reform in East Asia did not create food sovereignty; nonetheless, it did construct the juridical mechanisms needed to maintain the smallholder farming that is central to food sovereignty. At the same time it is important to emphasise that in East Asia pro-poor redistributive agrarian reform was about far more than land. Land reform, in the absence of a raft of additional measures that facilitate the capacity of petty commodity producers to increase their production, productivity and incomes, will not be beneficial to them, because they also need access to inputs at prices that they can afford, access to farm machinery, electricity and water at prices they can afford, credit at the right time and at the right price, and access to markets that pay prices that reflect their costs of production and not the prices that are set in global markets. Thus, pro-poor, gender-responsive redistributive agrarian reform requires extra supportive measures needed for farmers to improve the well-being of their families.[27]

Restricting land markets

While transformational food sovereignty movements are correct in advocating the need for pro-poor, gender-responsive redistributive agrarian reform, a second necessary condition is required if pathways towards food sovereignty are to be accelerated. Pro-poor, gender-responsive redistributive land and agrarian reform does not guarantee that the livelihoods of the rurally marginalised will improve. It does not do so because, following a reform, rural populations will continue to confront the reality of the market imperative, which requires that in the commodity economy of capitalism food providers must sell their products at competitive market prices if they are going to remain in business.[28] Market imperatives shape the operation of and the returns generated in local, regional, national and international markets. This in turn means that producers must continue to strive to be market-competitive, which requires continually striving to lower costs of production, which in turn requires that revenues from sales be directed towards investing in techniques and technologies that continually enhance market competitiveness. Not all producers will be able to meet the logic of the market imperative; indeed, this is a structural characteristic of capitalism. Those that are capable of meeting the market imperative will accumulate, while those that fail to meet the logic of the market imperative will turn to waged labour and to distress sales to meet short-term cash needs and then, later, to asset sales, including land. Eventually the market imperative differentiates producers into those who accumulate, innovate and expand and those who lose their assets and eventually have to rely upon selling their labour power for a wage in order to survive in the capitalist economy.

This has a critical implication for the outcome of pro-poor, gender-responsive redistributive land and agrarian reform: such reform provides a foundation for those who successfully meet the market imperative to acquire more land and other rural resources by using their accumulated surpluses to buy up the land and other rural resources of their neighbours who are consistently in deficit.[29] In other words, pro-poor, gender-responsive redistributive land and agrarian reform establishes the conditions by which land and other rural resources are later redistributed from the less successful to the more successful, and in so doing the pro-poor, and in all likelihood the gender-responsive, aspirations of the agrarian reform are negated. This means, in turn, that a successful pro-poor, gender-responsive redistributive agrarian reform requires restricting the market imperative

– most importantly, restricting land markets. Most centrally this would involve putting enforced legal restrictions on land transfers to ensure that agricultural holdings remained within a specified size distribution. Such restrictions are a central explanation of the East Asian development experience.[30] It would also be necessary to place enforced legal restrictions on the enclosure of common lands, which many smallholders rely upon to construct a viable livelihood. Cumulatively, while these restrictions would not de-commodify land, they would have an effect on the prices and quantities prevailing in land markets, as well as stabilising access to common property resources. This is a necessary condition of a successful pro-poor, gender-responsive redistributive agrarian reform, one which needs to be more forcefully articulated by practitioners of food sovereignty.

Three points need to be made about restricting land markets. The first is that the restriction of land markets is common in capitalist economies. Cadastral surveys define how land can and cannot be used, and there are limitations on the ease with which land defined for one purpose can be used for another. So, although restrictions on land markets may appear to be a radical departure from the tenets of neoliberal capitalism, such is not actually the case. The second point is that restrictions on land markets will require the intervention of a power capable of countervailing the strength of rural landed classes, and for better or worse that power must be the state – a point to which I will return later. Third, restrictions on land markets raise the question: what would be the incentive for those farmers who successfully meet the market imperative and are capable of sustained accumulation, innovation and expansion? Here, state fiscal incentives need to be created to push relatively more successful farmers into non-farm activities in order to continue to accumulate during the transition to a new food system. As was the case in parts of East Asia, when relatively more successful farmers are given economic incentives to diversify so as to sustain accumulation, diversification usually involves in the first instance the processing of agricultural output, which is more profitable than farming.[31] As successful farmers diversify, their need over time for their land diminishes and, while many successful rural enterprises continue to hold onto land for reasons of social protection and social status, the amount of land they hold on to is reduced as they continue to accumulate. This creates additional land availability for those who are not so successful, so that they can more fully utilise their available labour and non-labour resources to increase production and productivity. In this way, restricting land markets while providing the supportive measures necessary for smallholder farming to succeed can result in the emergence of livelihood-enhancing farm and non-farm economies sitting side-by-side in the countryside, a more prosperous rural economy that facilitates a reduction in rural poverty. While not underestimating the challenges facing the development of this kind of policy framework, these are not insurmountable, as this process has been witnessed in some cases of successful late industrialisation in East Asia.[32]

Agricultural surpluses

Restricting land markets can, in a sense, reduce the commodification of land and in so doing force more productive farmers into following other paths of accumulation. However, the key objective of land market restrictions is to facilitate increased agricultural surpluses among the more marginalised in the countryside rather than the relatively more prosperous. As is well known, farming has the capacity to produce more than those working in farming need to live and keep working; such 'agricultural surpluses' are the foundation of improvements in well-being.[33] A key objective of transformational food sovereignty movements, then, is the creation of a rural development framework that facilitates sustained increases in agricultural surpluses and manages them not on the basis of market demand, subsidies and dumping but on the basis of need, through supply management. Historically, of course, during the latter half of the 20th century the development frameworks that sought to sustain increases in agricultural surpluses were the technological traps of the Green and gene revolutions.[34] These must be forsaken in the quest for sustained increases in agricultural surpluses, and in favour of sustainable pro-poor, gender-responsive biotechnological change.

Sustainable pro-poor, gender-responsive biotechnological change is predicated upon maintaining rural environmental and natural resources.[35] The sustenance of soil integrity, the use of appropriate quantities of water at the appropriate time, seed sovereignty, sustainable cropping choices and patterns, and local and appropriate fertiliser, pest management and farm equipment technologies – all are central to sustainable biotechnological change in order to boost production, productivity, agricultural surpluses and incomes.[36] Where resources are degraded, sustainable pro-poor, gender-responsive biotechnological change requires the restoration of those environmental and natural resources. In order to do this, indigenous knowledge needs to be shared, particularly through farmer-to-farmer networks, as has been done across Central America, East Africa and Brazil.[37] A necessary correlate of sustainable pro-poor, gender-responsive biotechnological change in agriculture is the reassertion of agricultural research and extension as a public, not private, good, an end to the privatisation of agricultural research and extension, and the re-establishment of publicly funded and disseminated agricultural research and extension not directed towards the rurally prosperous, as was the case in the past, but instead towards meeting the livelihood challenges of the rurally marginalised.[38]

Agroecological farming

A critical part of meeting the livelihood challenges of the rurally marginalised is to facilitate the deepening and widening of agroecological farming practices, as is recognised by transformational food sovereignty movements.[39] This requires optimising the sustainable use of low-impact local resources and minimising the use of high-impact external farm technologies.[40] A correlate of such an agrarian strategy is that farm input and output choices, as well as local diets, need to be based, as they have been for all but the past century or so, far more on local ecologies, landscapes and ecosystems than on the needs of distant external markets.[41]

The benefits of an agroecological rural development strategy are several. First, agroecological practices are far more employment-intensive than industrial agricultural practices, and as such meet a key challenge of the twenty-first century: creating jobs.[42] These may not be the kinds of jobs that people would prefer to take, but for the underemployed relative surplus population such jobs are a vital part of the process by which their livelihoods are improved. As the East Asian case demonstrates, labour-intensive agriculture increases production and productivity as structural transformation occurs, and this would be an important part of an agrarian transition propelled by food sovereignty.[43] Moreover, even in the developed capitalist countries there are many who, if farming provided a living wage, would opt to attempt to farm out of choice. This is particularly so among younger people, seeking the autonomy of a sustainable lifestyle that farming has the potential to provide. Second, agroecological practices sustain soils and micronutrients and in so doing not only maintain the integrity of the soil but sustain and indeed improve its productive potential. This is of critical importance, because built into an agroecological rural development strategy must be the ongoing effort not only to sustain but in fact to increase crop yields by paying far closer attention to ecological requirements, input requirements, output choices and labour needs.[44] One of the foundational myths of the corporate food regime is that industrial agriculture is required to feed the ever-growing population of the world.[45] However, this is a myth. Granted, agroecological practices as they are currently constituted are not the dominant form of farm production around the world. However, copious scientific research worldwide indicates that agroecological production has the capacity to be as productive and profitable as industrial agriculture; indeed, once inter-temporal environmental impacts are included in the assessment of the costs and benefits of alternative farm production systems, agroecological production has the capacity to be more productive and profitable than industrial agriculture, while at the same time being more labour-intensive.[46] For example, it has recently been estimated that if world consumption of meat were halved the caloric 'savings' to food balance sheets would allow two billion people to be fed.[47] Agroecology, as a production system, is far more attuned to a nutrition-led farm production system than a market-led farm production system. As such it has the potential to supply the world, including those who are currently systemically food insecure, with a nutritious diet that not only generates jobs but also has far less impact on climate than the current industrial agriculture model. In other words, from the perspective of sustainable human well-being, labour-intensive, high-productivity agroecological production is a necessary component of twenty-first-century agriculture.

Local food systems

A shift to agroecological production systems brings with it an important correlate widely articulated by transformational food sovereignty movements: the need to build gender-responsive local food businesses, economies and systems.[48] For all but the past 150 years food systems have been local; while international trade in grain predates Roman times, the large-scale, long-distance movement of food has always constituted a small fraction of global production and as such control of trade was a foundation of empire.[49] The benefits of more localised food systems are several. For a start, historically, healthy local food systems are superior when it comes to distributing food to those most in need; localised mechanisms of social reciprocity that are central to moral economies ensure that, barring the impact of nature, members of a community are in receipt of a minimum standard of living, including access to food.[50] In this way healthy local food systems are far better at ensuring the health of communities – even communities riven by socioeconomic and political inequalities.[51] Healthy local food systems are also more resilient in the face of unforeseeable shocks, in that they are better at ensuring both that food is distributed to those in need of it and that production quickly returns to pre-shock levels.[52] Finally, local food systems have far less impact on the climate and are thus far more sustainable than food systems that rely upon the large-scale, long-distance movement of food.[53]

Having said this, the local should not be reified. Local food systems are sites of class, gender and racialised privilege, among others and, as a consequence, should not be viewed by definition as more equitable or more environmentally just. Rather, the issue is that local food systems that operate in conjunction with and reflect local landscapes,[54] because of both their relatively more manageable scale and the greater scope for localised action, are optimal sites upon which to accelerate progress toward a more just food sovereign system.

The state as a contested space

If the conditions necessary for constructing pathways that accelerate progress towards food sovereignty so far elaborated are relatively clear, they bring with them a condition that, in an era of neoliberal globalisation, may seem fanciful. Pro-poor, gender-responsive redistributive agrarian reform, restrictions on land markets, the fostering of sustainable gender-responsive biotechnological change and agroecological farming practices, and the resurrection of gender-responsive local food systems all require the intervention of a power capable of challenging capital and capitalism. Those conditions necessary to build food sovereignty involve heavy doses of new forms of regulation, interventions in the operation of 'free' markets, and challenges to the prevailing capitalist order. The power to undertake this range of interventions remains, for better or worse, the state.[55] While changes in the food system may be initiated from within communities and social movements, as is currently the case, such changes cannot be generalised without the involvement of a state that responds to the assertion of popular economic sovereignty by managing markets to the extent needed to tame capitalist impulses.

While the need for pro-poor, gender-responsive state intervention to transform the food system might be clear, the possibility of such an occurrence may appear to be wishful thinking. However, despite the origins, evolution and stark realities of the modern capitalist state, the state need not be treated as the exclusive tool of capital. Around the world the state is a contested space with which advocates of the marginalised and the marginalised themselves engage in an effort to transform it.[56] Indeed, this is recognised by transformational food sovereignty movements on those occasions when they negotiate with states that purport to be advancing food sovereignty; they are not seeking to get the state 'on side', but rather to transform the state, recognising that such a change is a process. Transformational food sovereignty movements must continue to engage with the state, both from within, to make claims on the state and initiate social, political and economic changes from within, as well as from without, to enforce claims that are made on the state and to ensure that there is no backtracking from any positive social, political and economic changes initiated.[57] One key avenue of engagement for transformational food sovereignty movements should be to campaign from within and without the state to improve the social wage, in terms of well-being, income, health, education and access to opportunities in the countryside, as part of a set of poverty-elimination policies. In poor urban areas the aim is to facilitate the capacity of consumers to pay more for their food.

Moreover, engaging with the state facilitates the capacity of food sovereigntists to learn how to transform it. However ineffective it has been to date, the inclusion of food sovereignty in the constitutions of Ecuador, Bolivia, Mali and elsewhere is a response to intense engagement by transformational food sovereignty movements. In addition, the revival of the CFS is an outcome of engaging with states – the revival would not have taken place without the pressure of transformational food sovereignty movements working with sympathetic state representatives and sympathetic representatives within the UN system. Finally, while transformational food sovereignty movements are, in general, less adept at working with municipal and regional states, particularly in the North, they could significantly enhance their capacities by learning from elements within some of the progressive food movements that do work with municipal, regional and state-level governments to improve the operation of local markets, protect food system workers' rights, render affordable the terms and conditions by which land can be accessed, and commence the labelling of genetically modified organisms. In turn, these progressive food movements can learn from transformational food sovereignty movements about how effective organisation can improve the ability to express agency, as well as the structural features of the capitalist food system.[58] In this way, within the interstices of the progressive elements of global food movements the practice of food sovereignty could become more firmly embedded.

Of course, in dealing with the state there is an eminent need for caution. Yet the postwar social democratic state demonstrated the extent to which the state can be pressured into making redistributive historic compromises that significantly improve livelihood security.[59] In an era of neoliberal globalisation this lesson cannot be forgotten; a minimalist objective of transformational food sovereignty movements should be, first, the re-establishment of the public sphere as a step in the reconstruction of a redistributive state, in order to facilitate more fundamental transformations of state structures and power. This kind of 'radical pragmatism' will for many be a highly contentious assertion,[60] but in the messy reality of the present the construction of real utopias requires working within the interstices of the present in order to forge a future. Indeed, this is what many in global food movements currently do.

When engaging with and seeking to transform the state, transformational food sovereignty movements need to be acutely aware of the level at which they are engaging: in cases where the state is subject to periodic elections it is more likely that claims can be made and changes initiated at the local level than at the regional or national level, if for no other reason than the fact that, in most instances, for voters electoral politics are primarily local. Indeed, one important if imperfect outcome to emerge from engaging with the state for progressive food movements – the establishment of food policy councils – has been a direct consequence of a focus on the local.[61] Granted, local states are more subject to class capture.[62] However, it is far more likely that a radical yet incremental act, such as the labelling of transgenic organisms, will be accomplished in a small US state or Chinese province, or indeed a mid-sized US or Chinese city, than in the country as a whole, as a first step. Yet that first step is of vital importance, both for its demonstration effects on neighbouring jurisdictions and for the effect that such a step would have on corporate behaviour, concerned as it is with maintaining access to markets.

Global trade

The requirement that transformational food sovereignty movements press for the reconstruction of a redistributive state is because for movement towards food sovereignty to be accelerated there is a need for a more interventionist state to initiate efforts to restructure global trade relations. It remains the case that the WTO, by regulating global agriculture, overseeing regional and bilateral free trade agreements, managing the trade-related investment measures that facilitate land grabbing, and in other areas, remains the most powerful global institution. In its activities, and as is rightly emphasised by food sovereigntists, the WTO is inherently neoliberal, promoting the ever-broadening and deepening of global capitalism through its principal agent, the TNC.[63] Therefore transformational food sovereignty movements have argued that agriculture should be removed from the purview of the WTO, in order to reduce the role of agro-food TNCs in the world food system. However, removing agriculture from the WTO does nothing to establish the kind of global trading arrangements that would facilitate acceleration towards food sovereignty. It would instead allow global markets to continue to operate in ways that benefit global TNCs.

A central demand of transformational food sovereignty movements must therefore go beyond limiting the role of TNCs by seeking to restrict the 'freedom' of global markets, so as to tame global market imperatives. This requires not so much the abolition of intervention in global markets as new forms of intervention that are more comprehensive – broader and deeper. The purpose of deeper intervention in global markets should be to reorient the purpose of trade away from the neoliberal objective of increased profitability and towards the more human-focused objective of improvements in well-being. In other words, the re-regulation of global markets should be done in order to transform food into what economists call a 'public good': something which is available to all and from which no one can be excluded.[64] Public goods are not immutable but are constructed through the struggles of citizens, including transformational food movements, which seek to disestablish the role of markets in social provisioning.[65] Thus, the transformation of food into a public good would go a significant way towards de-commodifying food and re-establishing the Polanyian idea that markets should be embedded in society rather than societies being embedded in markets.[66]

In order to re-regulate global trade towards meeting public needs, a pro-poor, gender-responsive state must lead efforts to replace the WTO with an International Trade Organization (ITO), as originally envisaged by John Maynard Keynes and Harry Dexter White. In the conception of Keynes and Dexter White, global trading arrangements following World War II were to be organised – and that is the word, organised – by the IMF, the International Bank for Reconstruction and Development, the International Clearing Union and the ITO, under the auspices of the UN General Assembly. The ITO was to be an institution that would facilitate economic and social progress by managing international trade cooperation between countries. This was to be achieved by:

- governing markets and stabilising prices through tariffs, quotas, subsidies, the treatment of skills and technology as global public goods, the management of commodity trade, the management of foreign direct investment and the explicit prohibition of dumping;

in order to speed progress towards

- full employment rooted in socially-acceptable labour standards, enhanced value-added, improved wages and working conditions, and ensuring the viability of small-scale producers.[67]

Thus, and in line with the demands of transformational food sovereignty movements, the ITO would have sought to protect small-scale producers by globally managing the supply of food and agricultural products and, in so doing, the incomes of small-scale producers.[68]

The governing of global markets should be an explicit objective of transformational food sovereignty movements; indeed, I would argue that the construction of an ITO is a necessary condition of achieving food sovereignty. Yet transformational food sovereignty movements cannot construct an ITO; this can only be accomplished by a consortium of pro-poor, gender-responsive states committed to improvements in human security through employment generation, labour-intensive, high-productivity production, and access to public goods and social provisioning as a precondition of enhancing human well-being.

A new 'common sense'

Clearly, none of the aforementioned measures will come about without pressure from global civil society on the state and on the international development and financial institutions. Thus, as in all politics, for global food sovereignty to reshape the food system depends on the current relation of forces between these elements and the dominant power of capital. Here the terrain is tipped against what are currently diverse and divergent movements of food providers and food consumers, and in favour of corporate interests and capital, with substantial support from the neoliberal capitalist state.

The strength of the forces of capital lay, in the domain of food, in the ability of capital's 'organic intellectuals' to define 'common sense': 'a chaotic aggregate of disparate conceptions',[69] a set of attitudes, moral views and empirical beliefs reflecting an individual's concrete experiences in society but lacking consistency or cohesion. Capital's organic intellectuals – 'the thinking and organizing element of a particular fundamental social class...distinguished...by their...function in directing the ideas and aspirations of the class to which they organically belong'[70] – negotiate individual subjectivity by welding together 'dispersed wills' into a shared awareness and meaning, from which emerges consent for class power because the ideologies which arise from the mediation of experience 'have a validity that is "psychological"'.[71] Thus sustained reiterations become accepted as truths: that the world cannot feed itself without industrial agriculture, that industrial agriculture requires transgenic organisms, that private property is sacrosanct, that localised food systems and petty commodity producers are relics of pre-modernity, and that the entry of capital into the food system has increased availability, 'choice' and 'freedom'.

However, it is clear that, if attention is focused 'violently' on the 'discipline of the conjuncture', and thus focused on understanding what is specific and different about the present,[72] neoliberal globalisation and the establishment of the corporate food regime has already produced something new: the call for food sovereignty and the forging of global movements for it. So transformational food sovereignty movements and their organic intellectuals need to do what they have already been doing for more than a decade: use praxis and learning to relentlessly contest contemporary 'common sense' across a range of arenas in social life in an effort to construct a new 'common sense' that configures different subjects, identities, projects and aspirations, building unity out of difference. Thus, as Eric Holt-Giménez notes,[73] transformational food sovereignty movements must reshape the 'common sense' of those other, progressive and reformist, elements of global food movements, with different and contrasting ideological perspectives on capitalist development and thus different agendas. Transformational food sovereignty movements must seek inclusively to find common ground with progressive and reformist food movements, 'building the moral and intellectual hegemony necessary for...a broad social consensus' that welds together dispersed wills into a new 'hegemonic bloc' around food.[74]

One central aspect of food makes this project eminently feasible: the fact that food cuts across the narrowly defined and socially constructed identities that have undermined class-based politics in the past half-century. As everyone on the planet is a food consumer, food has the potential to facilitate the development of a new, globally recognised, inclusive universal subject if it is articulated in unique and specific ways.

The groundwork for the construction of a new, inclusive 'common sense' around food has already been established: transformational food sovereignty movements have been a leader in fostering concern, of admittedly differing degrees, among food consumers about the circumstances facing food providers, whether they be the 'family farmers' of Northern agrarian populism or the 'peasants' and 'workers' of places 'out there'. At the same time food providers are often aware that food consumers are sympathetic to the straightened material circumstances they face. The issue, then, is to find more common ground between the food system's producers and consumers and the global food movement's producers and consumers.

Here, it would appear that transformational food sovereignty movements need to further elaborate two clear aspects of the corporate food regime that can be turned into sources of a stronger claim for change. The first is that the temperate industrial grain–oilseed–livestock agro-food complex of the corporate food regime is centrally implicated in climate change and ecological degradation.[75] Food–climate and food–ecological degradation links need to be relentlessly emphasised by transformational food sovereignty movements when arguing that a new food system must be a source of climate restoration rather than climate degradation. Sharply intervening in the ongoing evaluations and deliberations of the extremely visible Intergovernmental Panel on Climate Change is but one example of a way of inserting food more prominently into efforts to achieve climate justice, bringing out the linkages needed to give wider traction to the idea and practice of food sovereignty than is currently the case.

Second, it needs to be emphasised that food not only transcends the narrowly defined, socially constructed identities of the corporate food regime but is centrally implicated in the glaring livelihood inequalities that define the current conjuncture. Thus there are food–class linkages. This may not appear obvious, especially to food consumers. While petty commodity producers continue to be widely found in developing capitalist countries, it is often argued that food and agriculture occupies a minor place in the employment pattern of the developed capitalist countries. However, consider this: while in 2006, of an employed labour force of almost 17 million in Canada, only 2.2% were directly working in agriculture, work in the corporate food regime is not in farming: it is primarily in service-sector jobs – restaurants, bars and caterers; corner stores and supermarkets; and wholesale food trade. This has interesting implications for those who are involved in transformational food sovereignty movements: the numbers of people who are employed in jobs related to the plethora of activities that are part and parcel of the corporate food regime are not directly known.

If the numbers of those employed within the food system were directly estimated, it would be established that there is in fact a shared interest between petty commodity producers in developing capitalist countries and workers in developed capitalist countries. Thus, aggregating industrial classifications that can be directly linked to the food system in the 2006 Canadian census demonstrates that 13.8% of all employed Canadians were employed in a food system-related activity in that year.[76] It needs to be said that this is a dramatic understatement of the actual number of those whose livelihoods depend upon the food system: it does not include educators, researchers, government civil servants, financiers, or logistics operators, among others, who might be employed in a food system-related activity. It is significant that many of those who can be counted under the current classifications as working within the food system are not unionised and work in lower-wage jobs, while many of those who cannot be counted under current classifications but are in fact working within the food system are unionised and are working in better-paid jobs. The implication is clear: under the corporate food regime food is a critical livelihood issue for far more people than food providers. So the character of the corporate food regime must become a central concern for classes of labour in both the South and the North; there is a shared livelihood issue rooted in the inequalities promulgated by the corporate food regime, and this needs to be a key dimension in trying to construct a new 'common sense' around food among a broad democratic alliance of citizens united for change. Sharply intervening in the ongoing activities of organised labour, from the perspective of establishing collective bargaining units within business sectors, and negotiating over both the terms and conditions of employment and health and safety, is but one example of a way of inserting food more prominently into debates around economic justice. So too would be intellectual and organisational efforts to demonstrate that what are perceived to be the mutually incompatible demands of workers, farmers and consumers operating in different food system activities are in fact a set of shared interests. It is by bringing out the livelihood linkages between those who work in different parts of the food system within which they are enmeshed as consumers that wider traction for the idea and practice of food sovereignty than is currently the case can be constructed.

Towards food sovereignty

The objective of transformational global movements for food sovereignty is a livelihood-enhancing, climate-friendly food system that does not exclude anyone from food because it is available to all as a fundamental human right. This article has discussed a number of pathways that could accelerate movement towards food sovereignty. The foundation would be pro-poor, gender-responsive redistributive land reform, accompanied by extensive restrictions on land markets and the promotion of surplus-generating agroecological farming directed towards localised food systems. The reconstruction of a redistributive state is necessary to this, as a means of restricting local and global land, labour and product markets, as well as providing public goods and access to adequate forms of social provisioning. Cumulatively food sovereignty requires challenging the class power that is expressed in and through the corporate food regime by constructing a broad democratic alliance of peasants, smallholders, fishers, indigenous peoples, urban workers and underserved food communities prepared to confront the power of capital in the food system by fostering alternative modes of organising production and consumption in ways that contain elements of de-commodification and the re-regulation of markets on the basis of public need. The foundation of such a challenge must be the construction of a new common sense around food production and food consumption, which would elaborate shared identities and common interests, and thus allow agrarian and non-agrarian citizens to fully claim their individual and collective rights by establishing notions of democracy rooted in democratic economies, social and ecological justice, and the need for harmony between humans and nature.

Acknowledgements

I would like to thank Jun Borras, Eric Holt-Giménez, Alberto Alonso-Fradejas and Todd Holmes, as well as the three anonymous peer reviewers, who significantly improved the quality of this article.

FOOD SOVEREIGNTY

Notes

1. McMichael, *Food Regimes.*
2. Weis, *The Ecological Hoofprint.*
3. Akram-Lodhi and Kay, "Surveying the Agrarian Question."
4. Akram-Lodhi, "Contextualising Land Grabbing."
5. Akram-Lodhi, "Land, Markets and Neoliberal Enclosure."
6. Akram-Lodhi, *Hungry for Change.*
7. International Planning Committee for Food Sovereignty, "Définition de la souveraineté alimentaire."
8. These five principles have been paraphrased from the original six at the suggestion of a reviewer.
9. This paragraph is derived from Holt-Giménez and Shattuck, "Food Crises," 114.
10. Holt-Giménez, "Food Security?"
11. Holt-Giménez and Shattuck, "Food Crises," 114.
12. This paragraph is a consequence of a comment from one of the reviewers.
13. Bernstein, *Class Dynamics.*
14. Akram-Lodhi and Kay, "Surveying the Agrarian Question."
15. Freire, *Pedagogy of the Oppressed.*
16. Food and Agriculture Organisation, "Final Declaration."
17. de Schutter, "From Food Security to Food Sovereignty."
18. Intergovernmental Panel on Climate Change (IPCC) Working Group II, "Chapter 7."
19. The reference to pro-poor redistributive agrarian reform being gender-responsive is only implicit in the written statements of the transformational food sovereignty movement, but is explicit in its practice.
20. Akram-Lodhi et al., *Land, Labour and Livelihoods.*
21. Borras, "Redistributive Reform?"
22. Borras, "Questioning Market-led Agrarian Reform."
23. Wood, "Peasants."
24. World Bank, *World Development Report 2008.*
25. James, *Gaining Ground?*
26. Studwell, *How Asia Works.*
27. Borras, "Can Redistributive Reform be Achieved?"
28. Wood, "Peasants and the Market Imperative."
29. Akram-Lodhi, *Hungry for Change.*
30. Studwell, *How Asia Works.*
31. Ibid.
32. Kay, "Why East Asia overtook Latin America."
33. Ghatak and Ingersent, *Agriculture and Economic Development.*
34. Buckland, *Ploughing up the Farm.*
35. Carrillo, *Technology and the Environment.*
36. Pretty, *Agri-Culture.*
37. Holt-Giménez, *Campesino a Campesino*; Wilson, "Irrepressibly towards Food Sovereignty"; and Peterson et al., "Institutionalization of the Agroecological Approach."
38. Wolf and Zilberman, "Public Science."
39. Altieri, "Agroecology."
40. Altieri, *Agroecology.*
41. Davis, *Late Victorian Holocausts.*
42. McKay, *A Socially-inclusive Pathway to Food Security.*
43. Studwell, *How Asia Works.*
44. Altieri, *Agroecology.*
45. Conway, *One Billion Hungry.*
46. International Assessment of Agricultural Knowledge, Science and Technology for Development, *Agriculture at a Crossroads.*
47. Cassidy et al., "Redefining Agricultural Yields."
48. Patel, *Stuffed and Starved.*
49. Fraser and Rimas, *Empires of Food.*
50. Scott, *The Moral Economy.*
51. Ó Gráda, *Famine*; and Allen, "Realising Justice."
52. Fraser et al., "A Framework."
53. Erickson, "Conceptualising Food Systems."
54. Barber, *The Third Plate.*
55. Magdoff et al., *Hungry for Profit.*
56. Borras, "State–Society Relations."
57. IPPC Working Group II, "Chapter 7."
58. This paragraph has benefited from the comments of an anonymous reviewer.
59. Jessop, *State Power.*
60. Akram-Lodhi et al., *Globalization.*

150

61. Roberts, *The No-nonsense Guide.*
62. Mungiu-Pippidi, "Reinventing the Peasants."
63. Rosset, *Food is Different.*
64. Saul and Curtis, *The Stop*; and Akram-Lodhi, *Hungry for Change.*
65. Wuyts, "Deprivation and Public Need."
66. Polanyi, *The Great Transformation.*
67. George, "Alternative Finances."
68. Shaw, *World Food Security.*
69. Gramsci, *Selections from the Prison Notebooks*, 422.
70. Hoare, "Introduction," 3.
71. Gramsci, *Selections from the Prison Notebooks*, 377.
72. Hall, "Gramsci and Us."
73. Holt-Giménez, *Food Movements Unite!*
74. Akram-Lodhi, "Peasants and Hegemony."
75. IPCC Working Group II, "Chapter 7."
76. Akram-Lodhi, "Who is Working?"

Bibliography

Akram-Lodhi, A. H. "Contextualising Land Grabbing: Contemporary Land Deals, the Global Subsistence Crisis and the World Food System." *Canadian Journal of Development Studies* 33, no. 2 (2012): 119–142.

Akram-Lodhi, A. H. *Hungry for Change: Farmers, Food Justice and the Agrarian Question.* Halifax: Fernwood Books, 2013.

Akram-Lodhi, A. H. "Land, Markets and Neoliberal Enclosure: An Agrarian Political Economy Perspective." *Third World Quarterly* 28, no. 8 (2007): 1437–1456.

Akram-Lodhi, A. H. "Peasants and Hegemony in the Work of James C. Scott." *Peasant Studies* 19, nos. 3–4 (1992): 179–201.

Akram-Lodhi, A. H. "Who is Working within the Corporate Food Regime?" Haroon's Devlog. Accessed August 26, 2013. http://aharoonakramlodhi.blogspot.ca/2013/06/who-is-working-within-corporate-food.html.

Akram-Lodhi, A. H., and C. Kay. "Surveying the Agrarian Question (Part 2): Current Debates and Beyond." *Journal of Peasant Studies* 37, no. 2 (2010): 255–284.

Akram-Lodhi, A. H., S. M. Borras, and C. Kay, eds. *Land, Labour and Livelihoods in an Era of Globalization: Perspectives from Developing and Transition Countries.* London: Routledge, 2007.

Akram-Lodhi, A. H., R. Chernomas, and A. Sepehri, eds. *Globalization, Neo-conservative Policies and Democratic Alternatives: Essays in Honour of John Loxley.* Arbeiter Ring: Winnipeg, 2004.

Altieri, M. *Agroecology: The Science of Sustainable Agriculture.* 2nd ed. Boulder, CO: Westview Press, 1995.

Altieri, M. "Agroecology, Small Farms and Food Sovereignty." *Monthly Review* 61, no. 3 (2009): 102–113.

Barber, D. *The Third Plate: Field Notes on the Future of Food.* New York: Penguin Press, 2014.

Bernstein, H. *Class Dynamics of Agrarian Change.* Halifax: Fernwood Publishing, 2010.

Borras, S. M. "Can Redistributive Reform be Achieved via Market-based Voluntary Land Transfer Schemes? Evidence and Lessons from the Philippines." *Journal of Development Studies* 41, no. 1 (2005): 90–134.

Borras, S. M. "Questioning Market-led Agrarian Reform: Experiences from Brazil, Colombia and South Africa." *Journal of Agrarian Change* 3, no. 3 (2003): 367–394.

Borras, S. M. "State–Society Relations in Land Reform Implementation in the Philippines." *Development and Change* 32, no. 3 (2001): 545–575.

Buckland, J. *Ploughing up the Farm: Neoliberalism, Modern Technology and the State of the World's Farmers.* Halifax: Fernwood Books, 2004.

Carrillo, J. *Technology and the Environment: An Evolutionary Approach to Technological Change.* Instituto de Empresa Working Papers Economia 04-02. Accessed August 16, 2013. http://ideas.repec.org/p/emp/wpaper/wp04-02.html.

Cassidy, E. S, P. C. West, J. S. Gerber, and J. A. Foley. "Redefining Agricultural Yields: From Tonnes to People Nourished per Hectare." *Environmental Research Letters* 8, no. 3 (2013): 1–9.

Conway, G. *One Billion Hungry: Can We Feed the World?* Ithaca, NY: Cornell University Press, 2012.

Davis, M. *Late Victorian Holocausts: El Niño Famines and the Making of the Third World.* London: Verso, 2000.

Erickson, P. "Conceptualizing Food Systems for Global Environmental Change Research." *Global Environmental Change* 18, no. 1 (2008): 234–245.

Food and Agriculture Organization. "Final Declaration of the International Conference on Agrarian Reform and Rural Development." Accessed August 12, 2013. http://www.nyeleni.org/IMG/pdf/2006_03_FinalDeclaration_FAO_Conference_En-1-3.pdf.

Fraser, E., and A. Rimas. *Empires of Food: Feast, Famine and the Rise and Fall of Civilizations.* New York: Free Press, 2010.

FOOD SOVEREIGNTY

Fraser, E., W. Mabee, and F. Figge. "A Framework for Assessing the Vulnerability of Foods Systems to Future Shocks." *Futures* 37, no. 6 (2005): 465–479.

Freire, P. *Pedagogy of the Oppressed*. London: Bloomsbury Education, 2000.

George, S. 2007. "Alternative Finances." *Le Monde Diplomatique* (in English), January.

Ghatak, S., and K. Ingersent. *Agriculture and Economic Development*. London: Wheatsheaf, 1984.

Gramsci, A. *Selections from the Prison Notebooks*. London: Lawrence and Wishart, 1971.

Hall, S. "Gramsci and Us." In *The Hard Road to Renewal: Thatcherism and the Crisis of Left*, 161–173. London: Verso, 1988.

Hoare, Q. "Introduction." In A. Gramsci, *Selections from the Prison Notebooks*. London: Lawrence and Wishart, 1971.

Holt-Giménez, E. *Campesino a Campesino: Voices from Latin America's Farmer to Farmer Movement*. Oakland, CA: Food First, 2006.

Holt-Giménez, E, ed. *Food Movements Unite! Strategies to Transform our Food Systems*. Oakland, CA: Food First Books, 2011.

Holt-Giménez, E. "Food Security, Food Justice or Food Sovereignty?" Food First *Backgrounder* 16, no. 4 (2010). Accessed August 26, 2013. http://www.foodfirst.org/sites/www.foodfirst.org/files/pdf/Food_Move ments_Winter_2010_bckgrndr.pdf.

Holt-Giménez, E., and A. Shattuck. "Food Crises, Food Regimes and Food Movements: Rumblings of Reform or Tides of Transformation?" *Journal of Peasant Studies* 38, no. 1 (2011): 109–144.

International Assessment of Agricultural Knowledge, Science and Technology for Development. *Agriculture at a Crossroads: Synthesis Report*. Accessed August 26, 2013. http://www.unep.org/dewa/agassessment/re ports/IAASTD/EN/Agriculture%20at%20a%20Crossroads_Synthesis%20Report%20(English).pdf.

Intergovernmental Panel on Climate Change Working Group II. "Chapter 7: Food Security and Food Produc- tion Systems." Accessed April 16, 2014. http://ipcc-wg2.gov/AR5/images/uploads/WGIIAR5-Chap7_FG Dall.pdf.

International Planning Committee for Food Sovereignty. "Définition de la souveraineté alimentaire." Accessed August 12, 2013. http://www.foodsovereignty.org/FOOTER/Highlights.aspx.

James, D. *Gaining Ground? Rights and Property in South African Land Reform*. New York: Routledge, 2007.

Jessop, B. *State Power*. Oxford: Polity Press, 2007.

Kay, C. "Why East Asia overtook Latin America: Agrarian Reform, Industrialisation and Development." *Third World Quarterly* 23, no. 6 (2002): 1073–1102.

McKay, B. *A Socially-inclusive Pathway to Food Security: The Agroecological Alternative*. International Policy Center for Inclusive Growth Policy Research Brief 23. 2012. http://www.ipc-undp.org/pub/IPCPolicyRe searchBrief23.pdf.

McMichael, P. *Food Regimes and Agrarian Questions*. Halifax: Fernwood Books, 2013.

Magdoff, H., J. Bellamy Foster, and F. Buttel, eds. *Hungry for Profit: The Agribusiness Threat to Farmers, Food and the Environment*. New York: Monthly Review Press, 2000.

Mungiu-Pippidi, A. "Reinventing the Peasants: Local State Capture in Post-communist Europe." *Romanian Journal of Political Sciences* 3, no. 2 (2003): 23–37.

Ó Gráda, C. *Famine: A Short History*. Princeton, NJ: Princeton University Press, 2009.

Patel, R. *Stuffed and Starved: The Hidden Battle for the World Food System*. 2nd ed. Brooklyn, NY: Melville House, 2012.

Petersen, P., E. M. Mussoi, and E. Dal Soglio. "Institutionalization of the Agroecological Approach in Brazil: Advances and Challenges." *Agroecology and Sustainable Food Systems* 37, no. 1 (2013): 103–144.

Polanyi, K. *The Great Transformation: The Political and Economic Origins of our Times*. Boston, MA: Beacon Hill Press, 1944.

Pretty, J. *Agri-Culture: Reconnecting People, Land and Nature*. Abingdon: Earthscan, 2002.

Roberts, W. *The No-nonsense Guide to World Food*. 2nd ed. Oxford: New Internationalist Publications, 2013.

Rosset, P. *Food is Different: Why We must get the WTO out of Agriculture*. London: Zed Press, 2006.

Saul, N., and A. Curtis. *The Stop: How the Fight for Good Food Transformed a Community and Inspired a Movement*. Toronto: Random House, 2013.

de Schutter, O. "From Food Security to Food Sovereignty: Implications for the EU." Accessed August 16, 2013. http://www.eesc.europa.eu/resources/docs/de-schutter.pdf.

Scott, J. *The Moral Economy of the Peasant: Rebellion and Subsistence in Southeast Asia*. Oxford: Oxford University Press, 1976.

Shaw, D. J. *World Food Security: A History since 1945*. London: Palgrave Macmillan, 2007.

Studwell, J. *How Asia Works: Success and Failure in the World's most Dynamic Region*. New York: Grove Press, 2013.

Weis, T. *The Ecological Hoofprint: The Global Burden of Industrial Livestock*. London: Zed Books, 2013.

Wilson, J. "Irrepressibly towards Food Sovereignty." In *Food Movements Unite! Strategies to Transform our Food Systems*, edited by E Holt-Giménez, 71–92. Oakland, CA: Food First Books, 2011.

Wolf, S., and Zilberman, D. "Public Science, Biotechnology, and the Industrial Organization of Agrofood Systems." *AgBioForum* 2, no. 1 (1999): 37–42. Accessed August 26, 2013. https://mospace.library.umsys tem.edu/xmlui/bitstream/handle/10355/1230/Public%20science%20biotechnology%20and%20the%20indus trial.pdf?sequence=1.

Wood, E. M. "Peasants and the Market Imperative: The Origins of Capitalism." In *Peasants and Globalization: Political Economy, Rural Transformation and the Agrarian Question*, edited by A. H Akram-Lodhi and C. Kay, 37–56. London: Routledge, 2008.

World Bank. *World Development Report 2008: Agriculture for Development*. New York: Oxford University Press, 2007.

Wuyts, M. "Deprivation and Public Need." In *Development Policy and Public Action*, edited by M. MacIntosh and M. Wuyts, 13–39. Milton Keynes: Open University Press, 1992.

We are not all the same: taking gender seriously in food sovereignty discourse

Clara Mi Young Park[a], Ben White[a] and Julia[b]

[a]International Institute of Social Studies, The Hague, Netherlands; [b]Kalimantan Women's Alliance for Peace and Gender Justice, Pontianak, Indonesia

The vision of food sovereignty calls for radical changes in agricultural, political and social systems related to food. These changes also entail addressing inequalities and asymmetries of power in gender relations. While women's rights are seen as central to food sovereignty, given the key role women play in food production, procurement and preparation, family food security, and food culture, few attempts have been made to systematically integrate gender in food sovereignty analysis. This paper uses case studies of corporate agricultural expansion to highlight the different dynamics of incorporation and struggle in relation to women's and men's different position, class and endowments. These contribute to processes of social differentiation and class formation, creating rural communities more complex and antagonistic than those sketched in food sovereignty discourse and neo-populist claims of peasant egalitarianism, cooperation and solidarity. Proponents of food sovereignty need to address gender systematically, as a strategic element of its construct and not only as a mobilising ideology. Further, if food sovereignty is to have an intellectual future within critical agrarian studies, it must reconcile the inherent contradictions of the 'we are all the same' discourse, taking analysis of social differences as a starting point.

Introduction

The vision of food sovereignty (FS) calls for radical changes in agricultural, political and social systems related to food.[1] At the core this alternative vision is a 'radical egalitarianism',[2] a call for equality based on social change and social justice. This change also entails addressing inequalities and asymmetries of power in gender relations.

While women's rights are seen as central to FS, given the key role women play in food production and procurement, food preparation, family food security and food culture,[3] the way in which gender inequalities ought to be challenged and women's rights affirmed is unclear. FS itself, as Patel notes, is 'an intentionally vague call'; 'a call for people to figure out for themselves what they want the right to food to mean in their communities'.[4] This means that FS in general, and a gender perspective on FS in particular, are both difficult notions to pin down in empirical work. Few attempts have been made to integrate gender systematically in FS analysis.

The commitment to gender equality was asserted by La Via Campesina (LVC) at its fifth international conference in 2008, and reconfirmed at the sixth conference in Jakarta in 2013.[5] Women within LVC have actively advocated gender equality in programmes, manifestos and within the organisation itself,[6] and women's rights are considered of paramount importance for the realisation of FS.[7] Accordingly, the latest *Women of La Via Campesina Manifesto* (outcome of the 2013 LVC Jakarta conference) confirms gender justice and access to land as pillars of FS. The political project of the peasant women of LVC aims to fight neoliberal policies, capitalism and patriarchy towards the achievement of a 'new world' based on gender equality, social justice and food and seed sovereignty. As a feminist activist explains:

> Linking food sovereignty and feminism is therefore the unavoidable challenge facing social movements such as LVC, a challenge that implies reviewing their areas of focus and their strategies with a view to advancing gender-equal ways of working and the empowerment of women.[8]

However, the focus in FS discourse on 'the convergence of interests of groups who live on the land' means that class and other divisions among the rural poor may be ignored or downplayed.[9] The problem is that, when gender interests are incorporated in FS discourse, the same essentialising, 'we are all the same' discourse tends to appear again,[10] this time in relation to peasant or rural women. Therefore, for example, while prioritising family farming together with gender equality, FS does not systematically address the issue of intra-household inequalities and women's unpaid labour.[11]

> Our biggest step towards ending injustice in the world is taken by breaking the poverty cycle and granting the rightful place that *we peasants* have to provide and guarantee sufficient and balanced food for the peoples, recognizing the central role of women in food production... *We women* demand a comprehensive Agrarian Reform to redistribute land with our full participation and integration throughout the process, ensuring not only access to land, but to all the instruments and mechanisms on an equal footing, with a just appreciation of our productive and reproductive work, where rural areas guarantee a dignified and fair life for us.[12]

In this article we attempt to strip both food sovereignty and gender equality of their advocacy rhetoric, in order to highlight some underlying notions and issues and some areas where further theorising and empirical work may be needed. We use case studies of corporate agricultural expansion in five African and Asian countries to highlight the different dynamics of incorporation and struggle on the ground, depending on women's and men's different position, class and endowments. These, in turn, contribute to processes of social differentiation and class formation, and thus create rural communities more complex and antagonistic than those sketched in FS discourse and based on populist claims of rural egalitarianism, cooperation and solidarity.

The subjects of these case studies – like many rural people worldwide – are involved in a process of agrarian change, which reduces their involvement in autonomous food production and increases their engagement with corporate agriculture. They are not involved in FS movements, and they are probably unaware of FS discourse, as such. But they are the kinds of people and communities that FS, whether as vision of a desired future or as analytical frame, needs to encompass in its orbit.

We first need to deal with some of the problems involved in trying to make analytical sense of gender issues within FS discourse. This means finding ways to recognise and problematise, rather than side-stepping them, the dualisms and diversities in contemporary 'farming' and 'agriculture'; the heterogeneity of rural 'communities' and of 'rural women'; and the gendered relations in which they are involved.

Starting points

Our starting-point is that the potential strengths of FS discourse lie in 'the heuristic approach to power relations that it invites, particularly with respect to gender'.[13] FS is about 'power and control in the food system';[14] that is where we should explore the links between gender and food, highlighting gender dimensions of the (lack of) control over food systems in which people are involved, whether as producers or consumers.

While local circuits of agrarian self-sufficiency, and new 'nested markets' linking producers and consumers exist and may expand,[15] most crop producers live wholly or partly by supplying the food and other needs of urban and/or distant non-cultivators and are dependent on various kinds of larger-scale, corporate actors to make these links.

Here it is useful to adopt Bernstein's distinction between *farming* – 'what farmers do', production on the land and 'their social and ecological conditions and practices, labour processes and so on' – and *agriculture*, a much broader notion embracing 'farming together with all those economic interests, and their specialised institutions and activities, "upstream" and "downstream" of farming that affect the activities and reproduction of farmers'.[16]

In fact, virtually all common crops entering wider (national and global) circuits – including those on which the second part of our paper focuses – can be efficiently farmed (cultivated), with high per-hectare yields, on small-scale farm units. However, they require larger-scale units to take care of downstream (and in many cases upstream) activities. Cultivators are therefore, like it or not, engaged with the corporate sector in one form or another in what are often complex commodity chains.[17]

Small-scale farming communities, once involved in commodity production, experience chronic tendencies to processes of internal class-like differentiation. However, for various reasons these only rarely result in the later shift from 'differentiation' to (class) 'polarisation' as predicted by some rigid Marxist formulations. There are many possible counter-tendencies, some of them demographic or generational (for example, the splitting up of larger holdings among many children), some of them the result of the inherent resilience of 'peasant' farming,[18] some even the result of public intervention through progressive land taxation and/or land redistribution.

Besides these class-like dynamics and tensions, '"community" and its reproduction are always likely to involve tensions of gender and intergenerational relations'.[19] Understanding and analysing the dynamics of gender inequality on the one hand, and the differences that exist among women on the other, should be one central pillar of FS analysis. Consideration of gender in FS discourse means considering problems in gender relations not only on (small-scale) farms, but also in different positions in agrarian labour regimes,[20] and at different points in agro-commodity chains. A democratic conversation about food and agriculture policy must involve women positioned in many different capacities in food systems,[21] including: as direct producers on their own account, or as unpaid family labour in family farms; as direct producers on the land/farms of others (wage workers); as actors (own-account, unpaid family workers, wage workers) in the corporate upstream and/or downstream entities in agro-commodity chains; as providers of care and food, in reproductive roles; and as consumers of food and other agricultural products which they have not themselves produced. To assume shared experiences, shared interests and shared struggles among women in all these (often overlapping) positions – as FS discourse often appears to do in advocacy mode – requires quite some discursive acrobatics.

Access to land is one of the pillars of food sovereignty. On average, women comprise less than 20% of all landholders in developing countries and, when they do control land, generally have plots that are smaller, and of inferior quality, compared with those of men.[22] Gender issues in access to land have featured prominently since 2002 under the Global Campaign for Agrarian Reform (GCAR), with activities spanning research to advocacy and awareness raising.[23] However, Monsalve questions reforms that aim to strengthen women's individual land property rights, indicating that these may actually be an element in a more general 'Trojan horse' of neoliberal agricultural and land policies based on individual land titling.[24] Along the same lines, writing about South Africa, O'Laughlin observes:

> A narrow emphasis on legalizing women's individual right to land embeds a standard neoliberal proposition – the centrality of privatization and the commodification of land – within the liberal language of human rights. It focuses our attention on gender inequality in inheritance of property, of which the rural poor, women and men, have very little. Concerned with securing the property of those who have, titling excludes those who have not.[25]

Patriarchal power relations and gender inequalities in land rights, decision making and voice, among others, may have a decisive influence in incorporation in and exclusion from expanding corporate agriculture. These have been largely overlooked, as feminist scholars have noted,[26] in studies conducted from an agrarian political economy perspective. In focusing on class relations as the main unit of analysis, agrarian political economy has tended to ignore the role of other social relations in shaping relations of production and reproduction.[27]

Table 1. Case studies used for this article.

Country (region)	Crops introduced	Form(s) of incorporation
Ghana (Northern Region)	Mango	Wage labour and contract farming
Indonesia (West Kalimantan)	Oil palm	Wage labour and contract farming
Lao PDR (Provinces of Vientiane Capital, Vientiane and Borikhamxai)	Jatropha Bananas Tobacco	Wage labour and contract farming
Tanzania (Northern Tanzania)	Flower and vegetable seeds	Wage labour and contract farming
Zambia (Southern Province)	Sugar cane	Contract farming

Sources: FAO, *The Gender and Equity Implications (Northern Ghana)*; Julia and White, "Gendered Experiences of Dispossession"; Daley et al., *The Gender and Equity Implications of Land-related Investments*; Daley and Park, *The Gender and Equity Implications of Land-related Investments*; and FAO, *The Gender and Equity Implications (Zambia)*.

The male-headed peasant family was often assumed to be the basic unit of production.[28] Equally the realm of domestic unpaid labour, and of uncommodified work more broadly, has been largely disregarded by agrarian political economy thanks to 'its fixation with the sphere of commodities where value is realised'.[29] This has prompted a call to acknowledge uncommodified work, domestic institutions and social relations, going beyond analyses which see gender only in terms of its function to capital.[30] FS advocates the full recognition of women's contribution to food production and appreciation of their productive and reproductive roles. However, the extent to which gender inequalities and power relations within the household and the community shape the way in which women can exercise and balance these roles is not fully explored, especially from the perspective of the intersectionality of gender with other social differences.

Differences between and within rural communities in the gendered experience of incorporation are rarely fully explored in studies of the gender implications of corporate land-grabs, agro-fuels expansion, capitalist farming, etc. We include in this critique the various studies in which we ourselves have been involved, and which we will discuss in the rest of this article. These studies also do not fully explore women's reproductive role in the food system. Nevertheless, they do allow us to detect variations in processes of gendered social differentiation on the ground, especially in relation to incorporation and resistance to corporate agriculture.

We are not all the same

The case studies from which we have drawn examples are shown in Table 1.

The following sections draw on the five cases to explore differences between and within communities in women's changing experience of access to or control of land and other resources; access to income-generating opportunities; voice and participation in decision-making processes at the household and community level; the division of labour; and access to food and household food situation.

Access to land and other productive resources

Gender inequality in property relations has many implications in terms of FS, especially when communities engage with corporate agriculture in lieu of or along with (independent) small-scale farming. Women's land rights are not just about formal property rights. They often comprise nested, overlapping, sometimes conflicting rights held under different tenure systems (statutory, customary, religious), land use and decision-making patterns, inheritance and marriage rules and customs, as well as rights to access and use communal lands for collection of non-timber forest products, firewood and medicinal plants.[31] They may consist of secondary use and access rights granted through male relatives, and thus subject to change along with any changes in the conditions of men's rights over land.

One clear example is the changes in land control that occur when households are incorporated in corporate agriculture through contract-farming schemes. In Indonesia most indigenous Dayak communities in West Kalimantan, while denying women access to the community's political space, acknowledge women's right to land under customary tenure. However, the expansion of oil palm plantations in the area has undermined women's customary land rights through the formal system of smallholder registration based on 'Family Heads' (*Kepala Keluarga*), whereby the husband is designated as the head of the family and thus registered as the smallholder. When the conversion to oil palm was made and households surrendered (on average) 7 ha of land under customary tenure to get 2 ha of contract-farming land in statutory tenure, the new contract-farming plots were nearly always registered in the name of the male household head. So in a single bureaucratic stroke, women lost rights to land and produce.[32]

In an Iban Dayak community close to the Indonesia–Malaysia border,[33] aside from losing their customary residential land and being relocated into the plantation compound, households were also forcibly dispossessed of their agricultural lands (rice fields, fruit and rubber orchards) and access to rattan, which is one of the main livelihood sources of the community. After some time the women received a token amount of compensation for their agricultural lands, but this was unilaterally fixed by the company.

My four rubber orchards were flattened down to the ground. How should I eat now? (Mrs. Kenyalang)

The clearance [of paddy fields] was done early in the morning. Initially, it was said to be for road construction. [I] just took what was paid. If [the company] said it was three hectares, then it was three hectares. [I] got about four million rupiah as compensation, and it is not worth as much as [my] lost rice field. (Mrs. Menoa)[34]

The loss of their lands and other sources of livelihood left women in these households with no other livelihood option than to work for wages on the plantation. The women are mostly recruited as daily labourers, while the men have a wider range of jobs available for them, from daily labourer (usually harvester), permanent contract labour and, for some, higher positions such as public relations officer.

Not all of the cases we have analysed involved the corporate acquisition of large tracts of land and dispossession of local people. However, all had gendered implications with regard to changes in control of land and its produce brought about by the investments, particularly with respect to access to communal land and forest areas for the collection of non-timber forest products (NTFPs). For instance, before a Jatropha investment started in Vientiane province in Lao PDR in 2008, local women accessed the area, which was communal land, to collect

NTFPs for household consumption and sale. The land, which was formally classified as state forest, had previously been used for the shifting cultivation of upland rice until the government banned shifting cultivation and allocated the area to the village as communal land. Since the Jatropha plantations started, however, the availability of NTFPs has declined, with repercussions for food availability and women's income.[35]

Similarly, in the Ghana case, when a mango plantation was established, all trees on the land except three were felled to ensure effective pest and disease control. Thus people had to travel longer distances to collect firewood, fruits and nuts. Women in Gushie and Dipale reported having to walk 3 to 4 km to access fuel wood, fruits and nuts.[36] The company, a mango production and processing business established in 1999, operated a relatively small plantation comprising 150 hectares of land, mainly relying on outgrowers for sourcing mangos.[37]

In the Iban Dayak case mentioned above, because of the destruction of customary protected forests, people lost access to rattan, which was a major contributor to household income, particularly for landless households who were most dependent on extracting forest products. One female head of household explained that life was harder for her and her young son as she lost significant income from rattan and had to take up various kinds of other work, such as rubber sharecropping, collecting agarwood and wage work as log carrier in other areas. She has to work longer hours with less income compared to when she could collect rattan.[38]

Access to employment and livelihoods

Wage labour represents another form of incorporation in corporate agriculture. In the five cases, as expected, those who were landless or had less land were more likely to work as wage labourers on plantations and as casual labourers for other farmers. For instance, in Lao PDR it was mainly younger women and men from land-poor households and ethnic minorities who were recruited as casual labourers on the banana plantations of a British-owned company with operations in Borikhamxai province.[39] Similarly, in the flower and vegetable seeds venture in Tanzania, which had a nucleus farm in addition to outgrower operations, most of the divorced, separated and widowed female wage workers interviewed were no longer farming and had migrated to the area having lost access to land in their home villages and having heard that there was employment in horticulture. Sixty-one percent of the company's employees were women, hired as permanent, temporary or casual workers on the nucleus farm.[40] In this case satisfaction among women employees was relatively high; most women interviewed had been with the company for between three and 16 years and said they 'would like other companies to help fellow women and casual workers to get the same benefits as us',[41] indicating a positive attitude towards incorporation through wage employment. These women had apparently gone through a successful process of proletarianisation and said they were much better off now as, 'even if they don't have money for food, they can always get loans having a proper job'.[42]

Among tobacco contract farmers in Lao PDR there was a significant difference between emerging capitalist farmers, who were hiring in labour and renting additional land to cultivate tobacco,[43] and smaller farmers, who were struggling to make ends meet. In both groups equally, however, women lamented having to bear the heavy burden of additional production for the market.[44] While incorporation translated into higher incomes for male farmers, albeit with differences between the more and less well-to-do ones, for their wives it meant more work, but no increase in the money directly available to them. Although many male focus group respondents said that it was the women who managed the household budget, teasingly alluding to their 'wives' power', the fact that women did not participate in community decision making, training and extension opportunities suggests that there was no real equality within households,[45] and no gender-neutral path of social differentiation.

There were also a few female heads of household who were struggling to keep up with tobacco production as all their children had left for white-collar jobs in the cities and they could not afford to pay for casual labourers to help out. The household structure and the number of its active members thus had as much influence as access to land on the capacity to engage in commercial agriculture.[46]

We also found cases of successful female outgrowers who were, in some instances, more productive and wealthier than their male counterparts. A prosperous Zambian woman sugar cane farmer, one of only four members in her (50-member) farmer group, employed 25 irrigation workers. Besides sugar cane, she grew enough maize on her 1 ha residential allotment to feed her family for the whole year. She had built a three-bedroom house and servants' quarters, owned a car and had bought a 900 m^2 housing plot in nearby Mazabuka town, where she had started to build a house, planning to rent it out 'to diversify my risk portfolio'.[47] An extension worker attributed the success of female outgrowers in this project to various reasons, including their heavier reliance than men on paid casual labour, their use of incentives and their careful, hands-on management style, working together with the workers under their supervision.[48]

Voice and decision making

Lack of access to land and patriarchal power relations hamper women's participation in community decision making and interaction with investors. In the mango investment in Northern Ghana, women were excluded from community consultations over land leases with the mango producing company. As a consequence, they received no compensation and were left out of contract-farming opportunities, while losing access to the land they had been using.[49] Julia and White report similar difficulties faced by Indonesian Hibun Dayak women, who were excluded from the village consultations over the establishment of an oil palm plantation. Customarily women are not allowed to participate in community politics and 'the voice of the men was considered to be the unanimous voice of the villagers'.[50]

Maasai women in polygamous households in Lepurko, Northern Tanzania, each had their own assigned plot and, as a consequence, many of them had their own individual outgrower contracts with a well-established Dutch flower and vegetable seeds company. However, focus group discussions indicated that these women were not free to decide what to plant on their own fields and had been instructed by their husbands to sign the outgrower contracts. They also had to consult with their husbands on how to spend the cash income, which was largely used to pay school fees for their children.[51]

In contrast, in the Zambian case, focus group discussions held with male and female sugar cane outgrowers revealed that, in male-headed households with women as the main (registered) members of the scheme, women had more decision-making power over the use of the income than when the man was the registered member.[52] 'They believed that through their involvement they have challenged negative attitudes about their capabilities and their rights. They also felt they were able to articulate their problems and identify solutions when necessary, particularly when in group settings'.[53]

Similarly, in the Tanzanian case, the women largely indicated that they preferred having their own contracts. As one woman put it, 'having a contract in my name feels good as the contract gives me security. It is easier to get loans from friends because they know you have a contract...and you will get income'.[54] The women who did not have their own individual contracts said that they decided jointly with their husbands on matters relating to land use and income, while men in all the focus groups claimed they consulted with their wives on all decisions regarding land. As one young man in Mareu elaborated, 'land cannot be sold without asking the wives; if this happens the wife can take the husband to court'.[55]

Division of labour

In all the cases studied women had specific tasks in farming and agriculture, were responsible for collection of non-timber forest products, fuel wood and water and bore more or less entirely (sometimes with the help of children) the weight of care and domestic tasks. Wives of contract farmers had increased workloads in farming and agriculture but also in domestic activities, sometimes having to prepare meals for casual labourers who were hired to help with the investment crop, as was the case with tobacco farmers in Laos. However, in Lao PDR and Tanzania women in focus groups indicated a number of cascading benefits in terms of more money available for children's school fees, medicines, household appliances and buying meat.[56] It is also worth mentioning, though not systematically addressed in the case studies, that women tended to weigh their personal benefits against the overall well-being of their families, thus confirming that households and gender relations are sites of contradiction and resistance as much as of cooperation, sharing and mutuality.[57]

With the exception of the Tanzanian flower and vegetable seeds company, women wage workers tended to be in non-permanent, worse-paid jobs, which are often segregated by gender, task and crop; they were also absent or underrepresented in managerial or supervisory roles. On average they earned less than men. Women also remained responsible for most domestic work, so that wage work brought an increase in their overall workload. In the Tanzanian case some married women said they were stressed from having everything on their shoulders and no help from their husbands, who do not 'interfere' at all with household activities like cooking.[58]

Access to land is crucial not only for food collection and production, but also for women's chances to participate on their own account in contract farming. In the case of the mango company mentioned above, 149 (12%) of the

1200 registered outgrowers were women.[59] Among the factors contributing to women's involvement was the fact that the chiefs of two communities in Geshiegu-Karaga District, which account for over 50% of all female outgrowers, made sufficient land available for mango cultivation to any community member wishing to join the scheme, including women. Women in the Northern Region are less likely to access land, particularly for growing perennial crops considered as granting certain rights over the land on which they are grown. Men are reluctant to allow women's engagement in perennial crop cultivation, particularly unmarried women whose future husbands may also advance claims on the land.[60]

These few examples indicate that women may have a variety of different experiences with regard to changes in access to and use of land. Some women appeared to be more independent as a result of their engagement with corporate agriculture. Conversely, others were not in a position to decide autonomously how to use the land in the first place and whether or not to join a contract-farming scheme. Even when they were, they might not have the capital do to so. Another example from Tanzania illustrates this last point. In contrast to flower seed farmers, among whom women accounted for 22% of the contracted farmers, only 12% of contracted vegetable seed farmers were women at the time of the fieldwork. Vegetable seed production is more capital intensive than flower seed production, as it needs more fertilisers and more pesticides, as well as irrigation. Most male outgrowers had access to irrigation either in the form of piped water or drainage channels, as well as to tractors for preparing their land. They generally tended to be richer than flower seed outgrowers and had started vegetable seed farming on top of other income-generating activities. Some of them were businessmen or teachers, while others were commercial farmers who grew maize or other crops for sale. The few women vegetable seed outgrowers (five out of 41) were generally divorced or widowed, or had husbands who had other activities of their own.[61] Notably, all the outgrowers in the vegetable seed focus groups, with the exception of one woman, had rented in land. As noted above, vegetables require a lot of water so the fields need to have irrigation canals or be close to water. The men's fields, ranging from 1.5–3.0 ha, were considerably bigger than the women's, which were mostly smaller than 1.0 ha. The cost of renting in land for growing vegetables is twice as much as it is for growing maize – TSh40,000 (US$26) versus TSh 20,000 ($13) per ha at the time of the fieldwork. This creates a barrier for most women who tend to have fewer resources than men.[62]

Similarly, in Laos, the high start-up costs of tobacco farming made it out of reach of poorer farmers, particularly women. Most of the farmers who had joined got the needed start-up cash through bank loans.[63]

Access to food and household food situation

Turning to issues directly relating to FS, it is noteworthy that the Tanzanian flower seed outgrowers were cultivating flowers on land that was previously used for food crops. Some had been able to rent additional plots of land to continue growing maize, but women in Kiserian and Lepurko pointed out that for them cultivating flowers meant having less land for maize.[64] In addition, because tending flowers can take up to 12 hours a day during the growing season, women were working longer hours and had almost no time for other activities, sometimes having to skip meals. Nonetheless, most female focus group participants welcomed having a cash crop, which could be useful especially when maize harvests failed.[65]

In Zambia the sugar cane company gave 0.5–1 ha of marginal lands to its outgrowers as domestic plots. Women mainly used these plots for household food crops, because cash crops were more time-consuming, while men were more likely to grow cash crops. Thus, in terms of food availability, female-headed households seemed to be doing better, even though the overall work burden hampered their ability to produce and sell cash crops.[66]

In both the Lao PDR and Tanzania cases many of those who reported greater cash income from their engagement with corporate agriculture did not report that this had improved their household food situation.[67] In Tanzania 71% of those consulted reported an improvement in their cash income, while only 49% noted an improvement in their household's food situation.[68] In Lao PDR 90% of respondents said they were better off in terms of cash income, but only 52% said they were better off in terms of their food situation.[69] In this case the majority of negative responses were directly related to women's having lost access to NTFPs and to concerns about food provisioning and availability. Unlike the landless and land-poor plantation workers, most of those engaged in contract farming also had access to family land for rice growing, thus highlighting the close relation between land and the labour question.[70]

That new cash income does not automatically translate to better access to food is also suggested by the Indonesian case. In the Iban Dayak community women who lost their land have also become wage labourers on the oil-palm plantation.[71] In general the community is increasingly becoming dependent on cash to purchase food, and on the plantation as the main source of cash income. But women in different positions may see this in a different light. One woman noted:

> [We] produced our own in the past, then, one should obtain and produce by one-self in order to have them [the food]. Now, we must purchase in order to obtain food as lands have all been condemned. None can be planted anymore. Like it or not, it's only money that talks now.

Contrast this with the wife of the plantation's public relations officer, who valued the greater ease of life made possible by the new availability of purchased foods delivered to her door:

> In the past, if we wanted to buy fish or other types of meat, we had to travel to Seluas [the sub-district town]. Now, these things come by themselves [to the village]...really, people deliver them.

Women in small-scale farmer households also complained about the rising price of food and other goods, which reduced purchasing power even with higher cash incomes.

These examples indicate that processes of social differentiation are taking place, with different implications and outcomes for different people within communities and households. In terms of FS this suggests the need to investigate more closely the ways in which gender inequalities and the agrarian questions of land and labour or 'fragmentation of labour' are mutually constituted,[72] if land and food sovereignty are to be realistically advocated within a framework of gender and social justice.

Conclusions

As has been seen in the previous section, new classes of labourers and farmers are emerging in the five cases through wage work available on plantations and contract farming opportunities. Paths of differentiation and class formation are gendered and can be diverse depending on the relative status, the terms of incorporation and entitlements of different individuals and household members. So while some women value the opportunity to engage with agribusiness, for others it can mean more work and few benefits. Furthermore, some women do not have the capital and the resources needed to participate in the first place. Finally, for many women engagement with agribusiness can mean loss of access to communal land and forest areas for the collection of NTFP and thus of livelihood opportunities.

In terms of food sovereignty the case studies suggest that women's access to the means of production, particularly land, and control over what to produce are still largely constrained by patriarchal relations operating at community and household level. As Agarwal emphasises, 'a nod to gender equality is not enough. The problems that women face as farmers are structural and deep-rooted, and would need to be addressed specifically.'[73] At the same time there are also important differences in women's experiences both within and between communities, shaped by the perceptions and opportunities that different (classes of) women have vis-à-vis diverse forms of incorporation. So, for instance, Maasai women who were persuaded by their husbands to enter their own individual contracts but did not control the income deriving from the cash crop complained that growing flowers meant less land for maize. In contrast, female sugar cane contract farmers in Zambia were making a good living for themselves and their families and reinvesting in agriculture and beyond.

The case studies we have reviewed support the argument that 'women are not all the same' in terms of endowments, position within the household and community, and also in relation to what they may want and expect for themselves and their families by engaging or not with corporate agriculture. While agro-ecological principles and recognition of women's role in the food chain may point to valid alternatives to corporate agriculture, these alternatives need to address issues around gender inequalities, patriarchal relations and class-based differences, taking into account the diverse positions and roles of different groups (and women in different positions within those groups, as highlighted by intersectionality perspectives[74]), rather than conflating them analytically.

In summary, we hope in this paper to have provided support for three linked arguments. First, proponents of FS need to address gender inequalities systematically as a strategic element in its construct and not only as a mobilising ideology. Second, if FS is to have an intellectual future within critical agrarian studies, it will have to reconcile the inherent contradictions of the 'we are all the same' discourse,[75] taking the analysis of social differences, such as class, gender and ethnicity, as a starting point to challenge existing inequalities of power. Finally, the incorporation of gender interests in FS discourse requires that tendencies to generalisation must be corrected by recognition and exploration of differences in women's experiences, interests and responses both between and within communities. To assume that all rural women would choose (small-scale/family) farming as opposed to engagement with corporate agriculture is quite a leap of faith.

Taking gender seriously within the framework of food sovereignty also means locating gender firmly within FS's radical egalitarianist vision. Since land sovereignty is one of the pillars of FS and the two are so closely interrelated, this means that women should have exactly the same rights in land as men, without compromise, as well as the same rights to participate as full members in farmer groups, organisations, cooperatives, etc, the same rights in decisions on what crops to cultivate, the same rights to access services, instruments and mechanisms, and the same rights to decide whether and on what terms to engage with corporate agribusiness. This calls for FS to support women's rights to land holistically, with just appreciation of their productive and reproductive role, as well as to address seriously the inequalities that constrain their ability to exercise these rights.

Notes

1. Wittman et al., "The Origins," 4.
2. Patel, "Food Sovereignty," 670.
3. Wittman et al., "The Origins"; and Patel, "Food Sovereignty."
4. Patel, "Food Sovereignty"; and Patel, cited in Wittman, "Food Sovereignty," 92.
5. During the fifth conference, held 16–23 October 2008 in Maputo, Mozambique, LVC launched a world campaign "For an End to Violence against Women". Patel, "Food Sovereignty"; and Wittman, "Food Sovereignty." Five years later, in Jakarta, at the Fourth Women's Assembly of LVC, peasant women from

all over the world reasserted the commitment to end violence against women and discussed achievements and ways forward, including a strategy for reinforcing and giving continuity to the campaign against violence in all countries. LVC, "Women's Assembly Evaluates its Global Campaign."

6. Wittman et al.,"The Origins"; and Wittman, "Food Sovereignty". For an account of gender issues within LVC and its Global Campaign for Agrarian Reform, see Monsalve, "Gender and Land."
7. Wittman et al., "The Origins"; and Patel, "Food Sovereignty."
8. Caro Molina, "Feminism and Food Sovereignty."
9. Cousins and Scoones, "Contested Paradigms," 44.
10. Brass, *Peasants, Populism, and Postmodernism*, 314.
11. Agarwal, "Food Sovereignty," 1249.
12. La Via Campesina (LVC), *Women of Via Campesina International Manifesto*, emphasis added.
13. Patel, "Food Sovereignty."
14. Ibid., 1.
15. van der Ploeg, Jan Douwe, *The Art of Peasant Farming*, 130.
16. Bernstein, *Class Dynamics*, 65, 124; and Bernstein, "Food Sovereignty via the 'Peasant Way'," 1051.
17. 'Corporate' in this sense can refer either to capitalist firms, state-owned companies, non-profit organisations or farmer-owned cooperatives. See White et al., "The New Enclosures," 619.
18. As recently expounded by van der Ploeg, *The Art of Peasant Farming*.
19. Bernstein, "Food Sovereignty via the 'Peasant Way'," 1046.
20. Labour regimes are 'different methods of recruiting labour and their connections with how labour is organized in production (labour process) and how it secures its subsistence'. Bernstein, "Labour Regimes," 31–32.
21. Patel, "Food Sovereignty," 2.
22. FAO, *The State of Food and Agriculture*.
23. Monsalve, "Gender and Land."
24. This position emerged during the debates held at the GCAR international seminar on "Agrarian Reform and Gender", in 2003 in Cochabamba, which convened representatives of peasant, indigenous and human rights movements from 24 countries. Within a position that lobbies for communal forms of land tenancy, the participants also recognised the need to strengthen women's land rights in different tenure systems and 'not only as individual private property'. Ibid., 200. They asked: 'how secure can individual entitlements to lands for peasant women be when established in a context of privatization and economic liberalization policies that have already brought about dispossession and loss of land of many families and communities?' Ibid., 198.
25. O'Laughlin, "Gender Justice," 203.
26. For comprehensive discussions of feminist critiques of agrarian and peasant studies, see Razavi, "Engendering the Political Economy"; and Deere, "What Difference does Gender Make?"
27. Razavi, "Engendering the Political Economy."
28. Deere, "What Difference does Gender Make?," 63. Razavi, "Engendering the Political Economy", reminds us of feminist scholars such as Agarwal, Folbre, Hart, Kabeer and Whitehead, who have amply documented intra-household gender inequalities.
29. O'Laughin, cited in Razavi, "Engendering the Political Economy," 207; and O'Laughlin, "Gender Justice," 204. As O'Laughlin notes, 'the dynamics of capital accumulation also depend on how non-marketed work affects the real wage and the prices of commodities that enter the circuit of capital. Above all, from the perspective of labour a livelihood does not depend on wage income alone, for it includes the unmarketed labour of women, children and men.' O'Laughlin, "Gender Justice," 191.
30. Razavi, "Engendering the Political Economy," 222.
31. Behrman, Meinzen-Dick and Quisumbing, "The Gender Implications"; and Daley and Park, *The Gender and Equity Implications*.
32. Julia and White, "Gendered Experiences of Dispossession."
33. Information and quotes about the Iban Dayak come from unpublished research conducted by Julia in October 2011, commissioned by the Economic and Social Empowerment Commission of Pontianak Archdiocese.
34. One million Indonesian Rupiah was equivalent to about US$90 or €70 at the time of the fieldwork.
35. Daley et al., *The Gender and Equity Implications*, 31.
36. FAO, *The Gender and Equity Implications (Northern Ghana)*, 18. In addition to farming, traditional sources of livelihood in this region are fishing and livestock for men, and petty trading of foodstuffs, charcoal production, collection of firewood and picking of dawadawa and shea fruit from the wild for sale for women. Ibid., 15.
37. Ibid., 14.
38. See footnote 6.
39. Daley et al., *The Gender and Equity Implications*, 31. It is important, however, to emphasise that these people had been dispossessed of control over land not by processes ignited by capitalist ventures but by past decades of government resettlement and land reform policies, as highlighted in Lund, "Fragmented Sovereignty," 900.

40. Of the total workforce, 49% are permanent employees (staff), 32% are temporary (specific) workers who have annual renewable contracts and 19% are transient workers (casual labourers), who may be hired for one, two or three months. Female participation is higher among specific (81%) and transient (64%) as opposed to permanent workers (48%). Daley and Park, *The Gender and Equity Implications*, 24.
41. Ibid., 23.
42. From notes taken by Park in Arusha during focus group discussions with female wage workers, June 15, 2011.
43. Tobacco farmers in Pakse Village lived in an area with more fertile lands and easier access to water and roads. In contrast, those living in in Somsanook Village were more isolated and farther away from the water. Both villages are in Pakkading District, Borikhamxai Province.
44. Daley et al., *The Gender and Equity Implications*.
45. Ibid.
46. Ibid., 32.
47. FAO, *The Gender and Equity Implications (Zambia)*, 29.
48. Ibid., 26.
49. FAO, *The Gender and Equity Implications (Northern Ghana)*.
50. Julia and White, "Gendered Experiences of Dispossession," 1012.
51. Daley and Park, *The Gender and Equity Implications*, 20.
52. FAO, *The Gender and Equity Implications (Zambia)*, 27.
53. Ibid., 29.
54. Daley and Park, *The Gender and Equity Implications*, 20.
55. Daley and Park, *The Gender and Equity Implications*.
56. Ibid; and Daley et al., *The Gender and Equity Implications*.
57. Whitehead and Kabeer, *Living with Uncertainty*; Jackson, "Gender Analysis of Land"; Razavi, *Agrarian Change*; and O'Laughlin, "Gender Justice."
58. Daley and Park, *The Gender and Equity Implications*, 25.
59. This is a relatively high rate of female participation compared with other mango producer groups in Ghana. FAO, *The Gender and Equity Implications (Northern Ghana)*, 23.
60. Ibid.
61. Daley and Park, *The Gender and Equity Implications*, 22.
62. Ibid., 23.
63. Daley et al., *The Gender and Equity Implications*, 33.
64. Daley and Park, *The Gender and Equity Implications*, 19.
65. Ibid., 21.
66. FAO, *The Gender and Equity Implications (Zambia)*, 28.
67. Daley and Park, *The Gender and Equity Implications*; Daley et al., *The Gender and Equity Implications*.
68. Daley and Park, *The Gender and Equity Implications*, 35.
69. Daley et al., *The Gender and Equity Implications*, 35. Additionally, in Tanzania, 28% of respondents noted 'no change' in their cash income and 1% said they were 'worse off'. Daley and Park, *The Gender and Equity Implications*, 35. In Lao PDR 29% of focus group participants said there had been no change in their food situation and 22% indicated that they were 'worse off' than before.
70. Daley et al., *The Gender and Equity Implications*, 43.
71. From field notes taken by Julia in October 2011.
72. Bernstein, "'Changing before our very Eyes'."
73. Agarwal, "Food Sovereignty," 1255.
74. According to intersectionality perspectives, gender must be understood in the context of multidimensional, cross-cutting power relations embedded in social identities. Collins, *Black Feminist Thought*; and Shields, "Gender."
75. Brass, *Peasants, Populism, and Postmodernism*, 314.

Bibliography

Agarwal, Bina. "Food Sovereignty, Food Security and Democratic Choice: Critical Contradictions, Difficult Conciliations." *Journal of Peasant Studies* 41, no. 6 (2014): 1247–1268.

Behrman, Julia, Ruth Meinzen-Dick, and Agnes Quisumbing. "The Gender Implications of Large-scale Land Deals." *Journal of Peasant Studies* 39, no. 1 (2012): 49–79.

Bernstein, Henry. "Labour Regimes and Social Change under Colonialism." In *Survival and Change in the Third World*, edited by Ben Crow, Henry Bernstein, Lawrence Harris, Terry Byres, Diane Elson, John Humphrey, Hazel Johnson, Ed Rhodes, Mary Thorpe, and David Wield, 30–49. Cambridge: Polity Press, 1988.

Bernstein, Henry. "'Changing before our very Eyes': Agrarian Questions and the Politics of Land in Capitalism Today." *Journal of Agrarian Change* 4, nos. 1–2 (2004): 190–225.

Bernstein, Henry. *Class Dynamics of Agrarian Change*. Black Point, Nova Scotia: Fernwood Publishing, 2010.

FOOD SOVEREIGNTY

Bernstein, Henry. "Food Sovereignty via the 'Peasant Way': A Sceptical View." *Journal of Peasant Studies* 46, no. 6 (2014): 1031–1063.

Brass, Tom. *Peasants, Populism, and Postmodernism: The Return of the Agrarian Myth*. London: Frank Cass, 2000.

Caro Molina, Pamela E. "Feminism and Food Sovereignty: Oxfam Online Discussion Essay." 2012. http://blogs.oxfam.org/en/blogs/feminism-and-food-sovereignty.

Collins, Patricia Hill. *Black Feminist Thought: Knowledge, Consciousness, and the Politics of Empowerment*. 2nd ed. New York: Routledge, 2000.

Cousins, Ben, and Ian Scoones. "Contested Paradigms of 'Viability' in Redistributive Land Reform: Perspectives from Southern Africa." *Journal of Peasant Studies* 37, no. 1 (2010): 31–66.

Daley, Elizabeth, and Clara Mi Young Park. *The Gender and Equity Implications of Land-related Investments on Labour and Income-generating Opportunities: Tanzania Case Study*. Rome: FAO, 2012.

Daley, Elizabeth, Martha Osorio, and Clara Mi Young Park. *The Gender and Equity Implications of Land-related Investments on Land Access and Labour and Income-generating Opportunities: A Case Study of Selected Agricultural Investments in Lao PDR*. Rome: FAO, 2013.

Deere, Carmen Diana. "What Difference does Gender Make? Rethinking Peasant Studies." *Feminist Economics* 1, no. 1 (1995): 53–72.

Food and Agriculture Organization (FAO). *The Gender and Equity Implications of Land-related Investments on Land Access, Labour and Income-generating Opportunities: A Case Study of Selected Agricultural Investments in Zambia*. Rome: FAO, 2013.

FAO. *The Gender and Equity Implications of Land-related Investments on Land Access, Labour and Income-generating Opportunities in Northern Ghana: The Case Study of Integrated Tamale Fruit Company*. Rome: FAO, 2013.

FAO. *The State of Food and Agriculture 2010–2011: Women in Agriculture – Closing the Gender Gap for Development*. Rome: FAO, 2011.

Jackson, Cecile. "Gender Analysis of Land: Beyond Land Rights for Women?" *Journal of Agrarian Change* 3, no. 4 (2003): 453–480.

Julia, and Ben White. "Gendered Experiences of Dispossession: Oil Palm Expansion in a Dayak Hibun Community in West Kalimantan." *Journal of Peasant Studies* 39, nos. 3–4 (2012): 995–1016.

La Via Campesina (LVC). *Women of Via Campesina International Manifesto*. Accessed August 6, 2013. http://www.viacampesina.org/en/index.php/our-conferences-mainmenu-28/6-jakarta-2013/resolutions-and-declarations/1451-women-of-via-campesina-international-manifesto-2.

LVC. "Women's Assembly Evaluates its Global Campaign to 'Stop the Violence Against Women' and Makes Plans for the Future." Accessed August 22, 2013. http://viacampesina.org/en/index.php/our-conferences-mainmenu-28/6-jakarta-2013/1418-women-s-assembly-evaluates-its-global-campaign-to-stop-the-violence-against-women-and-makes-plans-for-the-future.

Lund, Christian. "Fragmented Sovereignty: Land Reform and Dispossession in Laos." *Journal of Peasant Studies* 38, no. 4 (2011): 885–905.

Monsalve Suárez, Sofía. "Gender and Land." In *Promised Land: Competing Visions of Agrarian Reform*, edited by Peter M. Rosset, Raj Patel, and Michael Courville, 192–207. Oakland, CA: Food First Books, 2006.

O'Laughlin, Bridget. "Gender Justice, Land and the Agrarian Question in Southern Africa." In *Peasants and Globalizations: Political Economy, Rural Transformation and the Agrarian Question*, edited by Haroon A Akram-Lodhi and Cristóbal Kay, 190–213. London: Routledge, 2009.

Patel, Raj. "Food Sovereignty." *Journal of Peasant Studies* 36, no. 3 (2009): 663–706.

Patel, Rajeev. "Food Sovereignty: Power, Gender, and the Right to Food." *PLoS Medicine* 9, no. 6 (2012). doi:10.1371/journal.pmed.1001223.

Razavi, Shahra. *Agrarian Change, Gender and Land Rights*. Oxford: Wiley-Blackwell, 2003.

Razavi, Shahra. "Engendering the Political Economy of Agrarian Change." *Journal of Peasant Studies* 36, no. 1 (2009): 197–226.

Shields, Stephanie. "Gender: An Intersectionality Perspective." *Sex Roles* 59, nos. 5–6 (2008): 301–311.

van der Ploeg Jan Douwe. *The Art of Peasant Farming: A Chayanovian Manifesto*. Black Point, Nova Scotia: Fernwood Publishing, 2013.

White, Ben, Saturnino M. Borras Jr., Ian Scoones, and Ruth Hall. "The New Enclosures: Critical Perspectives on Corporate Land Deals." *Journal of Peasant Studies* 39, nos. 3–4 (2012): 619–647.

Whitehead, Ann, and Naila Kabeer. *Living with Uncertainty: Gender, Livelihood and Pro-poor Growth in Rural Sub-Saharan Africa*. IDS Working Paper 134. Brighton: Institute of Development Studies, 2001.

Wittman, Hannah. "Food Sovereignty: A New Rights Framework for Food and Nature?" *Environment and Society: Advances in Research* 2, no. 1 (2011): 87–105.

Wittman, Hannah, Annette Desmarais, and Nettie Wiebe. "The Origins & Potential of Food Sovereignty." In *Food Sovereignty: Reconnecting Food, Nature and Community*, edited by Hannah Wittman, Annette Desmarais and Nettie Wiebe, 1–12. Halifax: Fernwood Publishing, 2010.

Land and food sovereignty

Saturnino M. Borras Jr.[a], Jennifer C. Franco[b,c] and
Sofía Monsalve Suárez[d]

[a]International Institute of Social Studies, The Hague, Netherlands; [b]College of Humanities and Development
Studies of China Agricultural University, Beijing, China; [c]Transnational Institute, Amsterdam, Netherlands;
[d]Land Programme, Foodfirst Information and Action Network-International, Heidelberg, Germany

Land and food politics are intertwined. Efforts to construct food sovereignty often involve struggles to (re)constitute democratic systems of land access and control. The relationship is two-way: democratic land control may be effected but, without a strategic rebooting of the broader agricultural and food system, such democratisation may fizzle out and revert back to older or trigger newer forms of land monopoly. While we reaffirm the relevance of land reform, we point out its limitations, including its inability to capture the wide array of land questions confronting those implicated in the political project of food sovereignty. Our idea of the land framework of food sovereignty, described as 'democratic land control' or 'land sovereignty', with working peoples' right to land at its core, is outlined, with a normative frame to kick-start a debate and possible agenda for future research.

Problematising the land dimension of food sovereignty

Food sovereignty, according to La Via Campesina (LVC), is 'the right of peoples to produce and consume safe and healthy food in sustainable ways in and near one's territory'. Not everyone must self-provision in food nor must farmers produce only food crops by this definition, because it accepts a food (trading) system based on diverse agronomic conditions across and within societies. Northern consumers can continue to drink their coffee and enjoy fresh bananas; rice cannot grow in all places in the world; some land is well suited to produce coconut and income from that farm enables the producer to purchase food produced by someone else. The 2007 Nyéléni meeting elaborated on this basic definition and brought forward the critical issues of land and land-based social relations in stating that food sovereignty 'ensures that the rights to use and manage our lands, territories, waters, seeds, livestock and biodiversity are in the hands of those of us who produce food. Food sovereignty implies new social relations free of oppression and inequality between men and women, peoples, racial groups, social classes and generations'.[1] Raising the land issue deepens the discussion by turning the spotlight on the presumed social foundations of food sovereignty and those who actually do the producing.

In this paper we use the term 'working peoples' (plural form) to refer to those who must work to earn a living. Working peoples are the 'sovereign' implicitly referred to in food sovereignty, and using the term asserts a popular, non-elite, non-statist take on the idea.[2] In locating food sovereignty in the inter-twined fields of production, circulation/trade and consumption using a political economy lens, the picture emerges of socially differentiated 'peoples' with diverse relationships to land and labour. The nature of their access to and control over land resources partly shapes their attitude to labour, including most crucially whether to hire their own labour out or to hire in the labour of others. It also shapes their attitudes to different types of land use: extensive or intensive, fluid or fixed boundaries, land as main resource or as key to other resources (water, forest), for example. In an abbreviated and crude manner we sketch key profiles of the peasantry as follows: poor peasants have small operations or too little land to cultivate, and so they tend to hire out their labour to others. Rich peasants have bigger capital operations or land to till, and so they have to hire in labour to work the land in excess of what their household labour can work. A middle farmer typically has just enough land to farm and/or operation to manage so that s/he does not engage in any significant hiring out or in of labour – but is always in danger of sliding down to becoming a poor peasant as a result of various factors, including market and ecological uncertainties. The meaning of land and/or farm operation size is relative to specific contexts: a middle farmer in Java, Indonesia may have one hectare of land, 5 mu (one-third of a hectare) in China, 30 hectares in Brazil, or 1000 acres in Canada. A landless labourer is one who has no land to work at all, and relies on selling her labour power in order to earn money to buy food. A sedentary farmer may engage in more intensive land use, compared to a pastoralist's idea of extensive land use and land boundary that is porous, fluid and flexible; many fishers are part-time farmers; many non-timber forest product gatherers are linked to land, but via different starting conceptions of land access and control than their settled counterparts.

These social groups, often collectively referred to as 'rural poor', have earned the right to be included in the food sovereignty definition of working peoples. Their ability to engage or not in food production, and if so, in what type of farming, fishing and livestock keeping systems and output market orientation, depends on the kind of access they have to which kind of land and related resources. This will determine their insertion into one, two or all three of the overlapping fields of production, trade and consumption. The nature of their insertion into these fields establishes (potentially competing) standpoints: a landless poor rural labourer who does not produce or trade food must look for food that is affordable, while a rich farmer who participates in all spheres can more easily buy food in the market, but also yearns for a higher farm-gate price for his own (food) produce.

Extending this complex lens to peri-urban and urban spaces reveals an even more complex picture. More people will need food who are increasingly separated and distanced from conventional rural-based food production and traditional food granaries. At the same time a vibrant increase in urban food gardening activities, both for commercial purposes and for household consumption, is unfolding in both Northern and Southern countries alike. There are various entry points of different social groups into the food sovereignty political project. While all social groups are necessarily involved in consumption, not all are involved in food production and/or trade. All groups may identify with the notion of food sovereignty, but not necessarily for the same reason, and there may be inherent tensions between them. For instance, the non-food producing consumer wing of the movement may emphasise the issue of affordability, while the producers emphasise the importance of 'socially just' remuneration that enables food producers to make a decent living. Meanwhile, an array of working peoples is involved in the trade side of food sovereignty. Petty food traders, market vendors, street food cooks and vendors, restaurant chefs, food servers, food transportation workers – the list goes on of non-food producers who, in diverse ways, are involved in getting food to consumers. In food sovereignty circles it is often assumed that bypassing merchants, peddlers and vendors to promote direct producer–consumer markets is the way to go forward. If food sovereignty has any hope as a political project, then local and direct producer–consumer markets are indispensable, not least because of their potential to challenge the industrial food system socio-culturally and politically.[3] The challenge of feeding the world in the twenty-first century should not be taken lightly, as more people now consume food than participate in food production. Given the proliferation of mega-cities, it seems inescapable that addressing this complex challenge will necessitate some sort of large-scale public orchestration that goes beyond the capacity of currently self-organised, scattered, localised direct farmers markets, thereby raising the question of the state, and ultimately, of sovereignty itself.

Scholars have recently begun to ask: 'who is the sovereign in food sovereignty?',[4] and to point out that there are multiple sovereignties in food sovereignty.[5] Some problematise the state as one among several sovereigns, while others, like Schiavoni, push the discussion further by problematising competing sovereignties involving state and non-state entities.[6] This brings us, in turn, to the issue of competing claims over resources[7]. How much of which natural resources can or should the state assert its authority over? The answers to such questions cannot be taken for granted. State claims often impinge upon long-standing assertions of local communities over contested natural resources, while policies crafted in distant national capitals, including agricultural, trade, taxation, subsidy, water or investment policies, affect the prospects of rural working people to access land and chart their own food production, trade and consumption arrangements. If the state allocates 25% of a country's best lands to large-scale commercial land investments, this necessarily affects the ability of agrarian subordinate classes to construct food sovereignty. If a government prioritises large-scale commercial agriculture and commercial–industrial hydro-power in water allocation, this necessarily affects smallholder farmers working to construct alternative farming and food systems.[8]

The dominant food system relies on industrial, mechanised, large-scale monocultures, as well as on non-industrial, small-scale farms for its consolidation and expansion, and continues to need land and water resources to sustain the accumulation process. The transformation of many food crops into 'flex crops and commodities' in recent decades has dramatically increased the system's demand for land. Attractive for the multiple-ness and/or flexible-ness of their end uses in today's global context, flex crops and commodities are those that can be produced as (or whose by-products can be transformed into) food, animal feed, biofuels and other industrial commodities – further integrating multiple sectors into industrial–financial conglomerates. Sugarcane, soya, palm oil, corn, cassava, sunflower, rape seed and even fast-growing industrial trees such as eucalyptus, acacia and pine, are important examples.[9] Unsurprisingly, as food empires continue to expand and to see phenomenal profits,[10] processes of enclosure accompany the expansion of flex crop and commodity production sites, thereby raising the issue of land.

The power and limitations of land reform for food sovereignty

Land reform and human rights in agrarian social movement narratives

Land and land politics – that is, the politics of who gets what rights and access to which land, for how long and what purposes, and of who gets to decide – are defining issues for peasant movements, past and present. LVC, the original spearhead of the political project of food sovereignty, places land politics at the heart of the food sovereignty project: for peasants, without land, there is no food sovereignty. In LVC's discourse the land dimension of food sovereignty has typically been framed in relation to poor peasants' struggles against large private landholdings (*latifundia*) and for equitable redistribution via land reform, where the latter includes public support services ('agrarian reform') to create a mass of farmers with sufficient land for economically viable full-time farming.[11] Some of the most influential founding groups in LVC include militant agrarian movements whose struggles centred around land, such as the landless movement MST in Brazil, KMP (Kilusang Magbubukid ng Pilipinas) in the Philippines, important Central American movements including the farmworkers movement ATC (Asociación de Trabajadores del Campo) in Nicaragua, and the SOC (Sindicato Obrero del Campo) in Andalusia.

This initial framing of food sovereignty's land dimension was consolidated with the establishment of an internal working group on land issues within LVC, with the Honduran and Nicaraguan movements coordinating and Rafael Alegría of Honduras, the first General Secretary of LVC, at the helm.[12] This bloc of forces succeeded in thwarting an early attempt by the Indian rich-farmer-based movement KRRS (Karnataka State Farmers' Association[13]) to prevent land reform from becoming the centrepiece of a global campaign. LVC officially launched the Global Campaign on Agrarian Reform (GCAR) in 1999–2000. GCAR was a broad alliance anchored by LVC that brought in close LVC allies such as the Foodfirst Information and Action Network (FIAN) and the Land Research and Action Network (LRAN). In contrast to its high profile international campaign against the World Trade Organization in the 1990s, the land campaign took the form mainly of national-level advocacy, until the World Bank began to promote its neoliberal version of land reform, so-called market-assisted land reform. Throughout its effective lifespan (2000–06) the GCAR focused on mobilising against market-assisted land reform and for a reaffirmation of conventional redistributive land reform. This culminated in Porto Alegre, Brazil in March 2006 with the Food and Agriculture Organization (FAO)-sponsored 'International Conference on Agrarian Reform and Rural Development' (ICARRD).[14] On the basis of the ICARRD's outcomes, the International NGO/CSO Planning Committee for Food Sovereignty (IPC), a broader social movement alliance of which LVC was a member, revitalised its working group on land and territory, and continued to deepen the multi-constituency dialogue started in Porto Alegre. This process fed into a historic gathering the following year – the Nyéléni Forum on Food Sovereignty held in 2007 in Mali – which explicitly built on the 'Land, Territory and Dignity' civil society organisation (CSO) Declaration of ICARRD and dedicated two out of the seven main themes of the Forum to the issues of (1) access to and control over natural resources (land, water, seeds, livestock breeds); and (2) sharing territories and land, water, fishing rights, aquaculture and forest use between constituencies.

This dialogue and collective construction of a land and natural resource perspective from the point of view of food sovereignty is ongoing. In countries such as Brazil, Colombia, Senegal and Indonesia social movements are converging to form cross-constituency platforms to struggle for land, natural resources and territory. This process has enabled agrarian social movements to participate in and influence the UN Committee on Food Security (CFS)/FAO Guidelines on Responsible Governance of Tenure of Land, Fisheries and Forests at the global level. Their own 'CSO Guidelines', presented as an input to the official process, are the most systematic elaboration on how to govern natural resources for food sovereignty. Concepts such as the 'shared and self-governed lands', 'water bodies', 'spaces and territories', 'collective rights over natural resources', and 'Free, Prior and Informed Consent' played a paramount role in this vision. This process has encouraged LVC, as an active member of IPC, to start reframing its land campaign as a 'campaign for land and territory'[15]. In some regions, like Latin America, the steady increase in the membership of indigenous peoples' movements in LVC, as well as the increasing interaction with strong movements leading land struggles at national level, as in the case of the Zapatistas and some of the Andean indigenous organisations, have also played a key role in this process of reframing. Apart from these processes LVC's reframing of the land issue has been encouraged by others as well, and by the increasing discussion of the need for recognition, protection and democratisation of land systems in the context of resurging enclosures. The strengthening of customary rights to land is prominent in the final declaration of the LVC conference 'Stop Land Grabbing Now!', held in Mali in 2011:

> In Mali, the Government has committed to give away 800 thousand hectares of land to business investors. These are lands of communities that have belonged to them for generations, even centuries, while the Malian State has only existed since the 1960-s. This situation is mirrored in many other countries where customary rights are not recognised. Taking away the lands of communities is a violation of both their customary and historical rights...We will develop the following actions:

Expand and strengthen our actions to achieve food sovereignty and agrarian reform, to promote the recognition of customary systems while ensuring the rights of women and to ensure the rights to land and natural resources of the youth.

One region where the issue of defence, strengthening and democratisation of customary land rights and systems is crucial for the land question and for food sovereignty is Africa. African rural social movements are currently engaged in manifold initiatives and struggles in order to safeguard their customary rights, more or less aware that social relations governed through customary tenure systems are not always democratic in terms of class, gender and ethnic dimensions. But how the new dimensions of territory and customary land rights (neither of which privileges Western individual private property notions of land control) relate back to LVC's main framework of land reform has not yet been dealt with concretely within LVC. Historically major tensions have existed between peasants and land reform, on the one hand, and indigenous peoples and ethnic groups and their interest in their ancestral territories and customary tenure systems, on the other hand. Land reform programmes have encroached into indigenous peoples' territories and affected the lives and livelihoods of forest dwellers and pastoralists as well.

Finally, the human rights framing of the land question and the momentum gained by LVC's work towards a UN Declaration on the Rights of Peasants and Other People Working in Rural Areas before the UN Human Rights Council are likewise important in the process of reframing the land pillar of food sovereignty.[16] Transforming a claim into a human rights claim is asserting its universality. The push for a human right to land and territory is one that can unite and be upheld by all social groups suffering from oppression, discrimination and dispossession with regard to land. At the same time, it asserts that all these groups have an inalienable right to land, and that the right to land of some cannot extinguish the right to land of others.

Rural social movements often deploy legal strategies as part of a broader political strategy in their struggles for land. Here too, the human rights framework is key in challenging other international legal frameworks that tend to work against the rural poor, or in defending local communities from abuses by international actors. The human rights framework can be used in a variety of ways, and depends on contextual factors like the recognition that it commands in the political and legal culture of a given country; the existence of alliances with human rights activists and law professionals able to understand the dynamics of social movements and of connecting law and politics; and the level of organisation of a movement and its ability to forge broad alliances with different social actors.[17]

The limitations of a land reform framework in the context of food sovereignty

Land reform was, and remains, a powerful policy framework and political platform: it demands redistribution of large private landholdings as a step towards social justice. In our view land reform is certainly a necessary, but not a sufficient component of a master-frame for the political project of food sovereignty for several reasons.

First, in aiming to redistribute large private agricultural landholdings to poor landless and near-landless peasants,[18] conventional land reform is a powerful coercive and corrective public policy that applies to political economic settings marked by the highly non-egalitarian distribution of land ownership, whether as a result of previous waves of land grabbing, colonial agricultural policy or neoliberal restructuring.[19] Yet land is clearly a compelling issue even outside such a setting. The main target of the recent wave of enclosures in Africa has been non-private land – roughly two out of every three hectares of land in that region. Forest dwellers live in communities that lie outside already privatised lands dedicated to agricultural commodity belts, while pastoralists are under growing pressure. States have historically been suspicious and hostile towards mobile or nomadic peoples, almost always trying to settle them in delimited territories. Societies marked by violent conflict and undergoing processes of peace building and conflict transformation are usually confronted by questions of land restitution. Indigenous peoples and other ethnic groups in the global South and North are under pressure and losing territorial access and control as a result of aggressive encroachment by extractive industries. Young prospective farmers or urban gardeners in rural, peri-urban or urban spaces who want land are often not able to gain access. Working peoples renting land, including in the North, often need some kind of tenancy and rent regulation. Fish workers' and artisanal fishers' search for control over aquatic resources often links them to some kind of land control issue as well.

Indeed, a wide range of social classes and groups urgently need land, but in contexts and circumstances outside the conventional parameters of land reform. Some of the most ardent proponents of food sovereignty come from these sectors, as a quick glance at existing food sovereignty alliances will reveal: IPC for Food Sovereignty, the US Food Sovereignty Alliance, Nyéléni Europe and the International Union of Food, Agricultural, Hotel, Restaurant, Catering, Tobacco and Allied Workers' Associations (IUF). Framing land for food sovereignty in conventional land reform terms is unlikely to have much relevance for such groups. Even when the concept is redefined and broadened ('agrarian reform') to encompass all types of land issues, it remains a stretch.

A closer look at the issue of land reform helps to reveal important contradictions in the food sovereignty narrative in relation to the state. From the beginning of the food sovereignty narrative there was an explicit preference for non-state actors, spaces and institutional processes linked to community, neighbourhood and household. The recent integration of food sovereignty concepts into official discourse, law and policy by some South American left-wing national governments (Venezuelan, Bolivian, Ecuadorian) has prompted a flurry of studies problematising state sovereignty in food sovereignty. A land perspective can contribute to this discussion, since many land spaces the food sovereignty movement aspires to transform or claim are either regulated or claimed by the state. Engaging the state is unavoidable when calling for land reform: only the state has the authority to carry out society-wide redistribution of large private landholdings. Yet land reform has contradictory implications for another key element in LVC's food sovereignty vision, which is 'peasant autonomy'. This is because land reforms, historically, have generally increased state control over peasant societies and spaces, whether intentionally or not. Despite a reluctance to recast private property arrangements through land reform, states have nonetheless, on occasion, chosen to undertake land reform, perhaps in part in order to pursue an agenda of capital accumulation and expansion of control over populations and spaces. Indeed, the subjugation and control of peasants have long been intrinsic to state-building processes, with land reform playing a crucial role. This is not to suggest that land reform politics are not contested, because they are in the broader context of dynamic state–society interactions where boundaries between state and societal groups are fuzzy and fluid.[20] Political contestation and contradictions surface in food sovereignty practice because, putting land under the control of food sovereignty practitioners – whether through large-scale public land reform, or 'by stealth' through individual incursion into state roadside and railway-side strip of land, or by a defiant neighbourhood invasion of idle and abandoned and bank-foreclosed private lots – will inevitably bring said practitioners face to face with the state, whether as regulator, arbiter or claimant of the same land. This in turn warrants giving more focused attention to challenging and potentially controversial questions about the nature of the state–society relationship in the context of land and food sovereignty politics.

Second, reading between the lines of various manifestos of agrarian movements, one current in the vision of food sovereignty is anchored on the strength of the modern-day Chayanovian[21] farmer (eg 'small farmers can feed the world'). This vision, and the actions it inspires, is necessary and important; it serves as an anchor, a beacon, a reference point, a foundation. Without this reference point there is no food sovereignty movement to speak off. Yet the actually existing political project of food sovereignty, midwifed by LVC, has since taken on a life of its own, extending today far beyond the original LVC ideal. The food sovereignty movement remains an essentially agricultural and rural mass-based one. But over time it has gained a wider following, bringing on board all kinds of rural working peoples, from poor farmer, middle farmer and rich farmer, to fisherfolk, pastoralists, indigenous peoples and landless agri- and aquaculture workers, to rural net food buyers as well. (Notably LVC itself has a mass base that is highly differentiated in similar fashion.[22]) The movement has also gained strategic ground among those living in urban and peri-urban spaces. Here, we find a diversity of militants dedicated to food sovereignty: food justice activists, urban gardeners, food store keepers, food vendors, net food buyers and various strands of community-supported agriculture initiatives (CSA), as well as food distribution councils, as in Venezuela's *comunas*. The list goes on. What emerges is a picture uncannily close to the vision projected at the Nyéléni 2007 gathering cited at the beginning of this paper.[23]

The actually existing food sovereignty movement is a *polycentric* one: that is, a movement with many centres of power, where interpretation and implementation of the common political project vary, perhaps even at times quite significantly. Indeed, the food sovereignty political project imagined by LVC at its formation in 1993 has become a truly multi-class coalition marked by multiple objective and subjective sub-alliances. The small farmer-centric/-led view and current – and the land reform ideal that is intrinsic to it – remains central, but is no longer the only strategic power bloc within the food sovereignty political project today. Non-peasant and non-rural, non-agricultural social groups have also contributed enormously to building the food sovereignty movement's political muscle. Their distinct perspectives on land cannot be subsumed under the middle peasant imaginary of land reform, but must be recognised on their own terms and given space in an adjusted framework.

Third, despite all the discussion about producer–consumer links, food sovereignty discourse and studies about it tend to remain focused largely on the production side, while the two other spheres of food politics, namely, trade and consumption,[24] remain relatively underexplored, as does the nature of the linkages between the three fields. Some scholars have begun to remedy this problem. Burnett and Murphy have noted the ambivalence and contradictions of the food sovereignty movement on the question of (international) trade.[25] Holt-Gimenez and Shattuck call attention to food justice movements, which in contrast to the food sovereignty mainstream, are much more concerned with the issue of who gets what kind of food and how, contextualised within class and race politics.[26] Agarwal emphasises the gender dimension in understanding politics in each of these fields and in tracking the links between these fields.[27] Bellows et al problematise the disconnect between nutrition, the right to adequate food and food sovereignty, which has led to flawed understandings of nutrition (over-medicalised, over-processed) and impeded an appreciation of nutrition in analyses of local food systems and cultural traditions – largely to the benefit of transnational corporations (TNCs); and they emphasise the need to address structural violence against women (such as child marriage of girls) in order to effectively overcome serious health problems such as stunting and chronic malnourishment over generations.[28]

Indeed, as these examples suggest, it is important to deepen our understanding of the links between these fields and between different movement strata especially because they may be marked not only by solidarities and synergies but also by tensions and contradictions. Whether and how the latter are understood and addressed (or not) will ultimately have an impact on the movement itself, shaping – or misshaping, depending on one's standpoint – its future character and direction. Similarly, a land reform-oriented framing of the land pillar in food sovereignty falls short of effectively capturing and expressing the defining moment that was Nyéléni 2007 in terms of what food sovereignty means, what the food sovereignty movement stands for, and who it represents.

Much food gardening in practice (urban, peri-urban, rural) is not intended to earn a living, but rather to partially contribute to meeting one's own household's food needs. The output produced via this channel will most probably never enter the official statistics on food production or food circulation. Even if it did, the non-peasant current of the food sovereignty movement and its aggregate food output would doubtless constitute only a small fraction of the aggregate food quantity produced by farmers worldwide. Yet the number of people involved and the political implications of their involvement in the food sovereignty movement cannot be taken for granted. Their political value almost inherently outshines their contribution to the aggregate food supply: they can transmit ideas of solidarity and struggle against the dominant food system; they can shape opinion and influence other opinion makers in and beyond their communities; they can amplify public awareness of problems associated with the dominant food system, such as health concerns, the ills of concentration of corporate wealth and power, and so on. Land issues also affect these actors, but of a kind that conventional land reform is likely to ignore. Their issues are more about house lots, house lots with food garden, food garden allotments, public community gardens, common gardens, access to public roadside strips of land, squatting on idle and abandoned lands, occupying and producing on bank-foreclosed lands, access to common community forestry, and so on – all of which have profound implications for the democratisation and localisation dimensions of food sovereignty.[29]

Fourth, preserving, strengthening or reforming and democratising customary tenure is a compelling challenge in many parts of the world, but the land reform framework is inherently ill-equipped to deal with it. African customary tenure systems are currently facing several challenges. In contexts of war, displacement and migration the authority of clan heads and traditional family structures has been eroding. At the same time increased economic interest in land is challenging customary systems in several ways. In many countries governments are taking advantage of formal state ownership or effective control of land (and underlying natural resources) to allocate much family and communal lands held and used under customary systems to foreign or domestic companies and investors. The vast majority of rural lands in Africa, especially unfarmed forests, rangelands and marshlands, remain under customary control. Such common resources are a major asset of many rural communities and are often the main source of livelihood for the land-poor and landless.[30] Investors are pushing to establish and strengthen clear property rights, for example by acquiring formal land titles that often conflict with and undermine customary systems. Customary institutions are under severe pressure to transfer land rights from local communities to powerful outsiders. It is not uncommon for those entrusted by customary institutions with authority to manage lands, such as the head of a family, clan or village, to abuse their power and sell or lease customary lands to middlemen and investors without sufficient prior consultation with village members. As mentioned above, restorative struggles, ie 'reclaiming' and protecting the customary tenure systems may also be problematic when they do not address the challenge to democratise these institutions in terms of class-, gender- and ethnicity-oriented social justice. In the contemporary context struggles in and in relation to customary tenure systems are better captured by interlinked principles around 'reclaim, protect, democratise'.

Towards a possible reframing

Land reform is a necessary but insufficient component of the land dimension of food sovereignty – for reasons discussed above. The initial thinking by Peter Rosset and his comrades (Shalmali Guttal, Indra Lubis, Faustino Torrez, Morgan Ody and Elizabeth Mpofu) in their attempt at analysing the evolution of the LVC land campaign – land reform *and* territory – is a crucial step forward in rethinking and reframing the land dimension of food sovereignty.[31] Our exploration attempts to build on this step, and on the belief that alternatives result from navigating continuously back and forth between what is ideal and what is do-able. Land reform may initially appear to be impossible, until implementation starts and it gains momentum and takes on its own dynamic. In building the land pillar, food sovereignty movements must reflect ground-level realities: their mass base and what they actually do and don't do, aspire to or not, taking actually existing complex and messy land-based social relations as the starting point. It is also important to emphasise the principle of working peoples' effective access to, control over, and use of land along the lines suggested by Ribot and Peluso and their notion of 'bundles of power'.[32] We propose a shift from the call for technocratic 'land governance', to a call for 'democratic land control'; and from a call for land reform, to a call for 'land reform *and* land sovereignty' – where land sovereignty is understood as firmly grounded in land reform.

The challenge is on two fronts. The first front involves tackling the mainstream reframing of the land issue along the lines of land tenure security and land governance, which have generally been aimed at promoting a Western-style private-land property system to facilitate private capital accumulation. In the land policy literature 'security' means providing, promoting and/or protecting property rights of exclusive owners and/or users of land. Usually it means individual and private property rights, including the right to alienate, for the purpose of commodification of land or transforming land into something marketable. Titles are the expression of this so-called security. But the problem with this notion of 'security' is that it can mean anything, pro- or anti-poor. It can mean the property security of big landlords living in the capital city, and relying on tenants or farm workers to make the land productive. It can also mean the property security of corrupt government officials who might seize vast tracts of lands in far-flung publicly owned areas through anomalous deals and for speculative purposes. Security in land property can also mean security of the banks that are selling credit for profit, and need collateral in case of payment default. This is not the kind of land security that sits well with food sovereignty.[33] The second front involves tackling the narrow and limited scope of conventional land reform for the various reasons elaborated in the preceding section.

'Democratic land governance' resonates with food sovereignty. It is a formulation that politicises and historicises land politics against land concentration and in favour of working peoples, by problematising who gets how much of which kind of land, how and why. Following Borras and Franco, 'democratic land governance' is understood here as a political process, often highly contested, where access to, control over and use of land is categorically biased in favour of non-corporate and non-elite social classes and groups;[34] it is inherently part of the broader and strategic challenge of democratising the state and society. It includes administrative and technical processes, such as efficient land records and titles. But it goes beyond these to address the need for land-based wealth and power (re) distribution, and to recognise working peoples' own decision-making processes in governing land. It recognises reformist contributions from both state and societal actors, and creatively combines formal and informal, official and non-official, and state and non-state institutions and processes. Democratic land governance implies and links action at all levels: local, national and even international.

All this we call 'land sovereignty', a term which, we feel, captures the essence of democratising land control in the context of democratising the food system. As an alternative conceptual framework and political platform, land sovereignty is: the right of the working peoples, rural and urban, to have effective access to, control over and use of land, and live on it as a resource, space and territory. Simply put, it is working peoples' right to land. The use of the term 'sovereignty' here sounds awkward, but there seems to be no better term to capture the essence of 'rural and urban working class people's effective access, control and use' and to link it to 'food sovereignty'. Land sovereignty has the potential to be a campaign framework for all working peoples who are confronted by some kind of a 'land question' – wherever they are, whatever they do – and to enable them to identify with and forge solidarity with others.

Although democratic land governance – or land reform *and* land sovereignty – in the context of a democratic food system could look different in different settings, it would have some defining features which build on the concept of a human right to land developed over time by the human rights organisation FIAN.[35] The human right to land is at the core of democratic land control. It is peoples' right – individually and in community with others – to effectively access, use and control land and related natural resources in order to feed and house themselves, and to live and develop their cultures and territories partly in a self-determined way where appropriate. It is a right that is limited to the use of land by communities and individuals for reproduction and commercial purposes. It is different from the right to private property, and does not primarily refer to a right to buy and sell land. The human right to land entails a geographic dimension that privileges the local, and does not condone control of far-away lands by absentee owners.

In this human rights context states have the obligation to respect, protect and fulfil the right to land. Under human rights principles states should not undermine existing access to and control over land by communities or individuals using it in the way described above. Revisions to existing civil codes and domestic property law (including those relating to international investment and investor protection) might be necessary in some settings, in order to fully recognise (and democratise) customary, ancestral and/or informal land rights and their governing systems and to overcome legal doctrines that are contrary to the basic idea spelled out above. Moreover, states should protect peoples' use of the land – and control over it – from interference by profit-seeking third parties: domestic capital, TNCs or foreign states. Finally, states should ensure a policy environment that allows people to make sustainable use of the land to feed themselves, and which allows them to decide how to develop their lands, taking into account the right to land of future generations. In fundamental ways the human right to land partly challenges the mainstream and dominant legal doctrines and frameworks that currently govern land. Colonial era laws have persisted into the twenty-first century, providing vast powers to nation-states to decide about land access and control, often undermining working peoples' interests. The human right to land offers a powerful legal tool to support the land claim-making by working peoples, rural and urban.

Building on this understanding of the human right to land, key features of democratic land control emerge. The first, and the most important, is to ensure working peoples' effective control of land in the construction of food sovereignty. This means always privileging the right of people who need access or protecting the right of people who already have access to a plot of land, whether full-time or otherwise, for (food) production for exchange or for their own consumption. The second is to recognise the diversity of social classes and groups working for and/or implicated in the multi-class, multi-ethnic political project of food sovereignty, in order to adequately address the resulting breadth and plurality of land questions. The breadth and diversity of the land sovereignty political project is at the same time a source of great strength and potential weakness, as the convergence also means internalising tensions and fault-lines. How to be inclusive but not populist, selective but not sectarian, while prioritising the working and oppressed social classes and groups will doubtless be a continuing challenge. But perhaps the most vibrant – and viable – social movements ultimately are those that confront such tensions, while exploring potential synergies within a multi-class coalition.

The third feature of democratic land control involves bringing the state back in using an interactive state-society framework.[36] Land sovereignty is in part a reaction against the idea of and quest for the most efficient economic (re)allocation of land via market forces. Because free market forces respond primarily to profit motivation, and are almost impossible to hold accountable, there is a need to 'bring the state back in', and with it the idea of nation-state sovereignty. Only the state has the authority to mobilise state resources to protect working peoples' access to land, to overcome resistance to redistributing large private landholdings, and to enforce compliance from social forces in society. But limiting our definition of sovereignty to the nation-state risks overlooking an array of dynamic social relations around land;[37] and relying solely on the nation-state to advance and protect the land interest of the working peoples ignores the contested and contradictory nature of state power. There are limits to bringing the state back in but these do not warrant shifting to the other extreme either – eg an overly purist, anti-state version of 'people-centred' (or 'community-led' or 'peasant-led') land control. Working peoples by themselves lack the authoritative power to frame and enforce rules and coerce compliance from competing social forces, which is needed to effect desired social change society-wide. The more promising way is via an interactive state–society approach, treating social forces within and between the state and society in relational context, and recognising that:

> Rights and empowerment do not necessarily go together. Institutions may nominally recognize rights that actors, because of imbalances in power relations, are not able to exercise in practice. Conversely, actors may be empowered in the sense of having the experience and capacity to exercise rights, while lacking institutionally recognized opportunities to do so. Formal institutions can help establish rights that challenge informal power relations, while those informal structures can also undermine formal structures.[38]

In the era of neoliberal globalisation, land sovereignty also implies a reordering of international relationships. In human rights terms this means that all states and international organisations must respect and protect existing land-based social relationships in other countries and effectively regulate TNC and business enterprises, the international financial system and the trade and investment regime accordingly.

The fourth feature is the use of multiple policies and land property regimes as the policy expression of land sovereignty: land reform certainly, and also land restitution, home lot allocation, forest land allocations, share tenancy tenure security, promotion of community land banks, enlargement of community common forest and parks, labour justice reforms, and so on. If the bottom line principle is for socially differentiated working peoples to have effective access, control and use of land, then the land property regimes will be necessarily diverse within and between communities from one society to another, depending on current structural and institutional situations conditioned historically. These can be communal, community, state, or private property rights, depending on the specific setting. In settings marked by a high concentration of private land-holdings, redistribution through land reform to create individual freehold private property of small plots for land reform beneficiaries makes a lot of sense. But land reforms can also spur change towards non-privatised property regimes and tenure systems (eg in Mexico and China). Meanwhile, effecting real access, control and use in some settings does not have to put an end to the property claims of landowning elites; this can be achieved by reforming the *terms* of tenancy or leasehold arrangements (as in West Bengal). Maintaining communal or community control and arrangements within patches of state or private land may also be a possible modality, as in Colombia's *reservas campesinas*, community land banking in Thailand, ancestral domain claims in the Philippines, or ethnic minority forestland claims in Vietnam. Land can remain formally owned and/or controlled by the state, but effectively accessed, controlled and used by working peoples (as in Vietnam and China). Land policy forms do not predetermine the character of land control by working peoples: there are customary land arrangements that are elitist and exclusionary, while there are private property rights arrangements that are truly pro-poor and democratic. If we take working peoples' democratic land control from this perspective, what we see is multiple and diverse forms of land-based social relations – relations thus requiring equally diverse formal and informal institutional arrangements to govern such social relations between and within the state and society.

Conclusion: uneven outcomes and future research

Public policies and political platforms, either by the state or community or social movements, once framed and passed, acquire a life of their own, often morphing into something different from the original idea and prompting realignments of relevant social forces. Some actors not involved at the beginning may be drawn in at a later stage. This is true whether for a state-centric policy like land reform, or a social movement platform like food sovereignty. This was the case of state land reform in Zimbabwe from 1980 to the present,[39] and in the Philippines, which was boycotted by all agrarian movements during the early period of 1988–92, only to be engaged by the latter from 1992 onwards – with far-reaching political and policy implications.[40] The case of food sovereignty is similar. While it started as a non-state, community-controlled initiative, it has clearly evolved into something reaching beyond that as well, with far-reaching implications for how we think about and construct food sovereignty.

Food sovereignty initiatives often exist outside the shadow of the state and corporate market, as seen in direct producer–consumer markets and other forms of community-supported agriculture. But scaling-up food sovereignty initiatives enough to allow a society-wide overhaul and reform of the food system will require interaction with the state and market – even if conceived from a Chayanovian standpoint of a notion of vertical integration. Ultimately, to have a strategic impact, the political project of food sovereignty will have to leverage broader state–society interactions in the direction of the desired interpretation and implementation. Yet actors and their influence exist unevenly across space, time and institutions, and the interplay of contending social groups and classes vying to shape the character, pace, scope and trajectory of food sovereignty makes the political project of food sovereignty fluid and dynamic, constantly evolving and being redefined, resulting in unexpected and uneven processes and outcomes across space and time. Similarly, who gets what rights and access to which type of land, for how long and what purposes are questions that are played out variably across space and time, depending on historically determined structural and institutional conditions and the alignment of social forces within the state and in society.

Yet democratic land control is among the key determinants of food sovereignty. It is essential to providing the material basis to jumpstart a food sovereignty initiative, and it influences and shapes the character and degree of autonomy of a prospective or actual producer and whether s/he can go into food production or can transition to a food sovereignty-inspired production system. The construction of a democratic food system is therefore inconceivable without its necessary material and political precondition: democratic land control. The relationship is nonetheless two-way: democratic land control may be effected in a society; but without a strategic rebooting of the broader agricultural and food system such occasional democratisation of land control can easily fizzle out and revert back to older or trigger newer forms of land monopoly, as witnessed in past cycles of land reform worldwide. How are today's struggles for democratic land control and food sovereignty interacting with one another in the real world? To what extent are ongoing political experiments and practices in twinning democratic land control and democratic food systems succeeding and how can these be scaled up? What is the nature of the role of the state and market in these experiments and practices? These are just some of the empirical questions that are emerging to be researched in the future, hopefully sooner rather than later.

Acknowledgements

This paper builds on a discussion paper by Borras and Franco, "A 'Land Sovereignty' Alternative? Towards a People's Counter-enclosure" published by the Transnational Institute (TNI) and based on ideas first sketched out in a Food First workshop in June 2012 in California. We thank the workshop participants, including Eric Holt-Gimenez, Sergio Sauer, Tanya Kerssen, Alberto Alonso-Fradejas and Annie Shattuck, for their helpful comments. That paper was further discussed at a conference of La Via Campesina in Indonesia in July 2012, and we thank the participants, including Henry Saragih, Indra Lubis, Shalmali Guttal and Peter Rosset for useful comments. Elements of the current version appeared in a paper presented at the 'Property Rights from Below' workshop held at the Massachusetts Institute of Technology, February 28–March 2, 2014. We thank the workshop participants for their comments, especially the organisers: Olivier de Schutter and Balakrishnan Rajagopal. For this present paper we thank Marc Edelman and two anonymous peer reviewers for their helpful review and suggestions.

Notes

1. LVC, cited in Patel, "Grassroots Voices"; and McMichael, "Historicizing Food Sovereignty."
2. The notion of 'sovereignty' in food sovereignty and the role of the state have recently attracted important scholarly deliberations. See Edelman, "Food Sovereignty"; Schiavoni, "Competing Sovereignties"; and Patel, "Grassroots Voices" for some of the key lines of argument.
3. For a related but broader treatment on food sovereignty and trade, see Burnett and Murphy, "What Place for International Trade?"
4. Edelman, "Food Sovereignty."
5. Patel, "Grassroots Voices"; and McMichael, "Historicizing Food Sovereignty."
6. Schiavoni, "Competing Sovereignties."
7. Wolford et al., "Governing Global Land Deals"; Franco et al., "The Global Politics of Water Grabbing"; and Rosset, "Rethinking Agrarian Reform."
8. Mehta et al., "Water Grabbing?"
9. Borras et al., *Towards an Initial Understanding*.
10. van der Ploeg, *Peasants and the Art of Farming*.
11. See Rosset, "Re-thinking Agrarian Reform"; and Monsalve Suarez, "Grassroots Voices."
12. Rosset, "Re-thinking Agrarian Reform"; and Monsalve Suarez, "Grassroots Voices."
13. See Assadi, "'Khadi Curtain'."
14. Borras and Franco, *Transnational Agrarian Movements*; and Monsalve Suarez, "Grassroots Voices." ICARRD represents a turning point in several ways. (1) It is the first time that peasants, indigenous peoples, rural workers, fisherfolk, pastoralists and other civil society organisations (CSOs) have come together to start jointly framing the land question. See International NGO/CSO Planning Committee for Food Sovereignty et al., "Agrarian Reform." (2) It is a turning point in breaking the global land policy hegemony of the World Bank, emphasising the outstanding role that agrarian reforms have to play in fighting hunger, and the need for a sustainable development model and respect for human rights. ICARRD was a unique experience enabling rural social movements and other CSOs to participate in the process of preparing and holding the conference on an equal footing with governments, and in a way that respected their autonomy. Monsalve Suárez, *The FAO*, 34–36. In a sense ICARRD became a forerunner of the current participatory processes at the UN Committee on Food Security (CFS).
15. Rosset, "Re-thinking Agrarian Reform."
16. Edelman and James, "Peasants' Rights and the UN System."
17. The effects of using the human rights framework vary and can range from contributing during land conflicts, changing the way conflicts over resources are framed, opening up space for policy dialogue centred on people's lives, fighting against agrarian legislation biased in favour of corporate interests and formulating alternative legal frameworks. The limitations of the human rights framework relate to the context in which this framework is applied but also to the current status of development of the framework itself. Both can change and it is up to each movement to pragmatically decide if the existing limitations can be overcome and if it is worth investing in it. Since law is one of the means *par excellence* of exercising power, any people's movement trying to change power relationships cannot avoid dealing with legal issues. For an agrarian movement, therefore, the question is not whether to use legal strategies but rather which legal strategy to use. Monsalve Suarez, "Grassroots Voices," 13.

18. Griffin et al., "Poverty and Distribution of Land."
19. This is textbook land reform. But as Franco, "Making Land Rights Accessible", argues, land reform laws are neither self-interpreting nor self-implementing, and the actual interpretation and implementation depend on the political interactions between state and social forces. In this process there are conservative land reform laws that can lead to outcomes that benefit the poor, while there are progressive land reform laws that can actually benefit the landed elite.
20. Fox, *The Politics of Food in Mexico.*
21. The notion of an ideal middle farmer who neither hires out nor hires in labour.
22. See Borras et al., *Transnational Agrarian Movements.*
23. Holt-Gimenez and Shattuck, "Food Crises"; Anderson, "The Role of US Consumers"; and Alkon, "Food Justice."
24. We include the question of 'nutrition' in our category of consumption.
25. Burnett and Murphy, "What Place for International Trade?"
26. Holt-Gimenez and Shattuck, "Food Crises." See also Bernstein, "Food Sovereignty via the 'Peasant Way'."
27. Agarwal, "Food Sovereignty."
28. Bellows et al., *Gender.*
29. See Robbins in this issue.
30. Alden Wily 2012.
31. Rosset, "Re-thinking Agrarian Reform."
32. Ribot and Peluso, "A Theory of Access."
33. The idea of generic 'land tenure security' – albeit *not* the mainstream private property-centric version – is useful as a framework for political struggles for those who seek democratic land control. In this context it is more useful to follow and build on the General Comment of the Committee on Economic, Social and Cultural Rights No. 4 on the right to adequate housing, as follows:
    Tenure takes a variety of forms, including rental (public and private) accommodation, cooperative housing, lease, owner-occupation, emergency housing and informal settlements, including occupation of land or property. Notwithstanding the type of tenure, all persons should possess a degree of security of tenure which guarantees legal protection against forced eviction, harassment and other threats. States parties should consequently take immediate measures aimed at conferring legal security of tenure upon those persons and households currently lacking such protection, in genuine consultation with affected persons and groups.See also the recent Guiding Principles on security of tenure for the urban poor recently issued by the Special Rapporteur on the Right to Housing, at http://daccess-dds-ny.un.org/doc/UNDOC/ GEN/G13/191/86/PDF/G1319186.pdf?OpenElement, accessed June 25, 2014. This legal development has allowed affected social groups to fight against forced evictions regardless of whether or not people hold any formal titles to the lands they are occupying and using. Security of tenure in this sense of 'securing land rights' (for working peoples), is fundamental to our proposed reframing and clearly fundamentally different from the technocratic approach of formalising land rights and promoting private property rights.
34. Borras and Franco, "Contemporary Discourses."
35. See Künnemann and Monsalve Suarez, "International Human Rights," 123–139.
36. See Fox, *The Politics of Food in Mexico.*
37. Scott, *Seeing like a State.*
38. Fox, *Accountability Politics*, 335.
39. Scoones et al., *Zimbabwe's Land Reform*; and Moyo, "Three Decades of Agrarian Reform."
40. Borras, *Pro-poor Land Reform*; and Franco, *Bound by Law.*

Bibliography

Agarwal, Bina. "Food Sovereignty, Food Security and Democratic Choice: Critical Contradictions, Difficult Conciliations." *Journal of Peasant Studies* 41, no. 6 (2014): 1247–1268.
Alkon, Hope Alison. "Food Justice, Food Sovereignty and the Challenge of Neoliberalism." Paper presented at the 'Food Sovereignty: A Critical Dialogue' international conference, convened by the Yale Program in Agrarian Studies, September 14–15, 2013.
Anderson, Molly. "The Role of US Consumers and Producers in Food Sovereignty." Paper presented at the 'Food Sovereignty: A Critical Dialogue' international conference, convened by the Yale Program in Agrarian Studies, September 14–15, 2013.
Assadi, Muzaffar. "'Khadi Curtain', 'Weak Capitalism' and 'Operation Ryot': Some Ambiguities in Farmers' Discourse, Karnataka and Maharashtra 1980–93." *Journal of Peasant Studies* 21, nos. 3–4 (1994): 212–227.
Bellows, Anne, Flavio L. S. Valente, and Stefanie Lemke, eds. *Gender, Nutrition and the Human Right to Adequate Food: Towards an Inclusive Framework.* New York: Routledge, forthcoming.
Bernstein, Henry. "Food Sovereignty via the 'Peasant Way': A Sceptical View." *Journal of Peasant Studies* 41, no. 6 (2014): 1031–1063.

Borras, Saturnino. *Pro-poor Land Reform: A Critique.* Ottawa: University of Ottawa Press, 2007.

Borras, Saturnino Jr., and Jennifer C. Franco. "Contemporary Discourses in and Contestations around Pro-poor Land Policies and Land Governance." *Journal of Agrarian Change* 10, no. 1 (2010): 1–32.

Borras, Saturnino Jr., and Jennifer C. Franco. *A 'Land Sovereignty' Alternative? Towards a People's Counter-enclosure.* TNI Agrarian Justice Programme Discussion Paper. Amsterdam: Transnational Institute, 2012.

Borras, Saturnino Jr., and Jennifer C. Franco. *Transnational Agrarian Movements: Struggling for Land and Citizenship Rights.* IDS Working Papers Series 323. Brighton: IDS, 2009.

Borras, Saturnino, Marc Edelman, and Cristobal Kay. "Transnational Agrarian Movements: Origins and Politics, Campaigns and Impact." *Journal of Agrarian Change* 8, nos. 2–3 (2008): 169–204.

Borras Saturnino Jr., Jennifer C. Franco, Ryan Isakson, Les Levidow, and Pietje Vervest. *Towards an Initial Understanding of the Politics of Flex Crops and Commodities: Implications for Research and Policy Advocacy.* Think Piece Series on Flex Crops and Commodities No. 1. Amsterdam: Transnational Institute (TNI), 2014.

Burnett, Kim, and Sophia Murphy. "What Place for International Trade in Food Sovereignty?" *Journal of Peasant Studies* 41, no. 6 (2014): 1065–1084.

Edelman, Marc. "Food Sovereignty: Forgotten Genealogies and Future Regulatory Challenges." *Journal of Peasant Studies* 41, no. 6 (2014): 959–978.

Edelman, Marc, and Carwil James. "Peasants' Rights and the UN System: Quixotic Struggle? Or Emancipatory Idea whose Time has Come?" *Journal of Peasant Studies* 38, no. 1 (2011): 81–108.

Fox, Jonathan. *Accountability Politics.* New York: Oxford University Press, 2007.

Fox, Jonathan. *The Politics of Food in Mexico: State Power and Social Mobilization.* Ithaca, NY: Cornell University Press, 1993.

Franco, Jennifer. *Bound by Law: Filipino Rural Poor and the Search for Justice in a Plural-legal Landscape.* Manila: Ateneo de Manila University Press, 2011.

Franco, Jennifer. "Making Land Rights Accessible: Social Movements and Political-Legal Innovation in the Rural Philippines." *Journal of Development Studies* 44, no. 7 (2008): 991–1022.

Franco, Jennifer, Lyla Mehta, and Gert Jan Veldwisch. "The Global Politics of Water Grabbing." *Third World Quarterly* 34, no. 9 (2013): 1651–1675.

Griffin, Keith, Azizur Rahman Khan, and Amy Ickowitz. "Poverty and Distribution of Land." *Journal of Agrarian Change* 2, no. 3 (2002): 279–330.

Holt-Gimenez, Eric, and Annie Shattuck. "Food Crises, Food Regimes and Food Movements: Rumblings of Reform or Tides of Transformation?" *Journal of Peasant Studies* 38, no. 1 (2011): 109–144.

International NGO/CSO Planning Committee for Food Sovereignty, La Via Campesina, Sofia Monsalve, FIAN International, Peter Rosset, LRAN, Saúl Vicente Vázquez et al. "Agrarian Reform in the Context of Food Sovereignty, the Right to Food and Cultural Diversity: Land, Territory and Dignity." 2006. http://www.nyeleni.org/IMG/pdf/landandfoodsov_paper5.pdf.

Künnemann, Rolf, and Sofia Monsalve Suárez. "International Human Rights andGoverning Land Grabbing: A View from Global Civil Society." *Globalizations* 10, no. 1 (2013): 123–139.

McMichael, Philip. "Historicizing Food Sovereignty." *Journal of Peasant Studies* 41, no. 6 (2014): 933–957.

Mehta, Lyla, Gert Jan Veldwisch, and Jennifer Franco. "Water Grabbing? Focus on the (Re)appropriation of Finite Water Resources." Introduction to special issue of Water." *Alternatives* 5, no. 2 (2012): 193–207.

Monsalve, Sofia. *The FAO and its Work on Land Policy and Agrarian Reform.* Working Paper 11.11.11. Amsterdam: Transnational Institute, 2008.

Suárez Monsalve, Sofia. "Grassroots Voices: The Human Rights Framework in Contemporary Agrarian Struggles." *Journal of Peasant Studies* 40, no. 1 (2013): 239–290.

Moyo, Sam. "Three Decades of Agrarian Reform in Zimbabwe." *Journal of Peasant Studies* 38, no. 3 (2011): 493–531.

Patel, Raj. "Grassroots Voices: Food Sovereignty." *Journal of Peasant Studies* 36, no. 3 (2009): 663–706.

Ribot, Jesse, and Nancy Peluso. "A Theory of Access." *Rural Sociology* 68, no. 2 (2003): 153–181.

Rosset, Peter. "Re-thinking Agrarian Reform, Land and Territory in La Via Campesina." *Journal of Peasant Studies* 40, no. 4 (2013): 721–775.

Schiavoni, Christina. "Competing Sovereignties, Contested Processes: The Politics of Food Sovereignty Construction." *Globalizations*, forthcoming.

Scoones, Ian, Nelson Marongwe, Blasio Mavedzenge, Jacob Mahenehene, Felix Murimbarimba, and Chrispen Sukume. *Zimbabwe's Land Reform: Myths and Realities.* Woodbridge, UK: James Currey, 2010.

Scott, James. *Seeing like a State: How Certain Schemes to Improve the Human Condition have Failed.* New Haven, CT: Yale University Press, 1998.

Van der Ploeg, and Jan Douwe. *Peasants and the Art of Farming: A Chayanovian Manifesto.* Halifax: Fernwood, 2013.

Wolford, Wendy, Ruth Hall, Saturnino M. Borras Jr., Ian Scoones, and Ben White. "Governing Global Land Deals: The Role of the State in the Rush for Land." *Development and Change* 44, no. 2 (2013): 189–210.

Contextualising food sovereignty: the politics of convergence among movements in the USA

Zoe W. Brent, Christina M. Schiavoni and Alberto Alonso-Fradejas

International Institute of Social Studies (ISS), The Hague, Netherlands

As food sovereignty spreads to new realms that dramatically diverge from the agrarian context in which it was originally conceived, this raises new challenges, as well as opportunities, for already complex transnational agrarian movements. In the face of such challenges calls for convergence have increasingly been put forward as a strategy for building political power. Looking at the US case, we argue that historically rooted resistance efforts for agrarian justice, food justice and immigrant labour justice across the food system are not only drawing inspiration from food sovereignty, but helping to shape what food sovereignty means in the USA. By digging into the histories of these resistance efforts, we can better understand the divides that exist as well as the potential for and politics of convergence. The US case thus offers important insights, especially into the roles of race and immigration in the politics of convergence that might strengthen the global movement for food sovereignty as it expands to new contexts and seeks to engage with new constituencies.

Introduction

In autumn 2008, as volatile food prices, combined with financial recession, rapidly drove up hunger rates, food movement leaders from across the USA converged in New York for an event entitled 'Step Up to the Plate: Ending the Food Crisis'. Standing on the stage together were, among others, urban food justice activist LaDonna Redmond, farmer leaders Ben Burkett and John Kinsman, farmworker activist Gerardo Reyes, and food worker union leader Pat Purcell. While the event brought attention to the pressing challenges of the food crisis, what made it historic on a number of levels was the message of unity it conveyed in the context of a food system marred by deep historic divides. Tackling the food crisis, starting at home, would mean doing the hard work to overcome these divides, and this would require understanding how they had come to be. Redmond captured this sentiment, emphasising that achieving a just food system was not about returning to some idyllic past. The US food system had been anything but just, as it had been built on the exploitation of Indigenous peoples, slaves and others, and continued to function through exploitation and oppression. Achieving a just food system, she explained, would involve building something radically different from what had existed in the past and at present.

Little did the organisers and participants know at that time, but this event was an initial step in a process of convergence that would lead to the founding of the US Food Sovereignty Alliance (USFSA) two years later. The founding of the USFSA was an attempt both to unify food and farm groups in the USA into a more cohesive and powerful movement and to situate this domestic movement within the broader global struggle for food sovereignty – defined as 'the right of peoples to healthy and culturally appropriate food produced through ecologically sound and sustainable methods, and their right to define their own food and agri-culture systems'.[1] While originally popularised in the 1990s by the global peasant movement, La Via Campesina, the concept of food sovereignty has increasingly become a rallying cry, strategy and proposal by social movements across the globe, now spanning well beyond La Via Campesina and its farming base.

As resistance to the advance of a corporate-dominated global agri-food system continues to grow, reaching into new geographic areas and new social groups, the increasing diversity of actors who have taken up the banner of food sovereignty poses additional challenges, as well as opportunities, for already complex transnational agrarian movements.[2] In the face of such challenges calls for convergence have increasingly been put forward as a strategy for building political power (see, for instance, *Food Movements Unite!*[3]). Desmarais empha-sises the importance of 'unity in diversity' to understand this convergence,[4] while Martínez-Torres and Rosset explain the practice of '*diálogo de saberes*' (dialogue among knowledges) within La Vía Campesina as a means of navigat-ing sensitive points of tension.[5] Of course, attempts at convergence are not with-out their challenges, and Edelman et al. ask if such practices will 'allow for a constructive interchange between the various social groups differentiated by their actually existing practices of food sovereignty? The answers to these questions are not obvious and will require careful empirical research and conceptual soul-searching.'[6] This is particularly the case as food sovereignty expands into new settings that dramatically diverge from the agrarian context in which it was originally conceived.[7]

Below we will explore the case of the USA where, despite its small farming population, there is a diverse array of movements comprised of many different actors across the food system. These include movements for agrarian justice, food justice and immigrant labour justice, among others. As each increasingly engages with food sovereignty, this opens up new possibilities for dialogue and convergence, while also raising challenging questions. The US case thus pro-vides the chance to explore how food sovereignty is both shaping and being shaped by other food and agrarian movements. With this analysis we offer preli-minary insights into some of the ways that convergence among movements is already taking place, and some of the barriers that exist.

Convergence in the US context

In examining the spread of food sovereignty in the USA we emphasise that it is not starting from a blank slate, but rather from a highly complex landscape of historically entrenched power structures and diverse forms of resistance, with particular histories and dynamics within and between them. First, the workings of the US brand of agrarian capitalism and corporate power are dominating features of the US food system, as has been well covered elsewhere.[8] Second, resistance takes many forms. While there is a tendency among both activists and scholars to refer to a 'US food *movement*', what exists in reality is a patchwork of different, contrasting, even competing efforts. Occupying the dominant mainstream are the largely white and middle- to upper-class consumers promoting 'voting with your fork'[9] and other forms of conscious consumerism that emphasise both individual choice and change through the marketplace.[10] Some have criticised these efforts as missing critical pieces of analysis, including issues of race and class, and thus serving to deepen the divides in an already divided food system.[11] Moreover, the focus on so-called 'foodie' consumers as key protagonists renders invisible a host of other actors – those most marginalised within the food system – who are also on the frontlines of resistance. These largely invisible movements are the focus of this piece, particularly the divisions and politics of convergence within and between them, as they shape and are shaped by the growing movement for food sovereignty.

As part of a recent surge in food sovereignty literature scholars are increasingly examining what this concept means, or could mean, for the USA. Anderson has provided a helpful overview of the current landscape of food sovereignty efforts in the USA, citing the need for food sovereignty to speak to consumers if it is to take off in a US context.[12] Dickinson and Anguelovski, among others, emphasise that this is particularly the case with regard to low-income consumers with limited options, who have been largely bypassed by the recent resurgence in local, sustainable foods (with some notable exceptions through the efforts of the food justice movement, described below).[13] Clendenning and Dressler look at the degree to which urban agriculture and other urban food initiatives might serve as building blocks toward food sovereignty in the USA,[14] while Roman-Alcalá looks at a particular case in the San Francisco Bay Area in which global food sovereignty framing is inspiring a local struggle over land.[15] For Alkon worker-led initiatives are among the most promising bridges toward food sovereignty in the USA,[16] while Minkoff-Zern and Mares et al. note the unique challenges, as well as opportunities, facing farmworkers *vis-à-vis* food sovereignty.[17]

As movements evolve, studies are increasingly pointing to the dialectical nature of convergence, recognising that food sovereignty is shaping food and agrarian movements in the US *and* vice versa. In this regard Figueroa, calls for a 'relational, historically and culturally grounded, "people-centered" approach'.[18] In examining the predominantly Black community of Chicago's South Side, she argues that survival mechanisms passed down from the Jim Crow era, such as collective food purchasing, are sowing the seeds for food sovereignty in the aftermath of slavery, migration and de-industrialisation. She thus makes the case that such contextually specific articulations of food sovereignty are relevant to the broader food sovereignty movement. We argue that this is similarly the case with resistance efforts for agrarian justice, food justice and immigrant labour justice, which are not only drawing inspiration from food sovereignty but helping to shape what food sovereignty actually means in a US context. By digging into the histories of these resistance efforts, we find that the US case offers important insights, especially into the roles of race and immigration in the politics of convergence, which might strengthen the broader movement for food sovereignty as it expands to new contexts and seeks to engage with new constituencies.

While US-based food sovereignty efforts are diverse, here we explore the USFSA as a key site of convergence.[19] This analysis is also inspired by Tarrow's work on social movements and contentious politics. Throughout we draw on the concept of 'frames of contention', the ideas that 'justify, dignify, and animate collective action'.[20] We also use Tarrow's notion of 'repertoires of contention – how demands are made, what types of activities are practised and how mobilisation is sustained in different contexts.[21] Because this article cannot sufficiently engage with all the diverse frames and repertoires of contention animating food and agrarian movements in the USA today, we have opted to focus primarily on efforts around racial justice, rooted in struggles for Black Liberation, which are gathering momentum in the USA as we write. Converging under the banner of #BlackLivesMatter, among others, such organising has emerged in the wake of increased attention to state violence against people of colour, perpetuated over generations. In the words of #BlackLivesMatter co-founder, Alicia Garza, these mobilisations are 'an ideological and political intervention in a world where Black lives are systematically and intentionally targeted for demise'.[22] We therefore consider it timely to examine the ways in which these efforts intersect with radical food movement work. However, space constraints do not allow us to explore many other important mobilisations, not least of which are the centuries of Native American struggles for sovereignty and their intersections with other food and agrarian movements. This is covered elsewhere,[23] but certainly merits further attention. The same is true for many other struggles shaping the landscape of resistance across the USA.

Agrarian justice

Often absent from mainstream food movement narratives is the fact that many family farmers are living in a state of crisis, with 30% of US farms having disappeared over the past 50 years.[24] Also overlooked is the significant role that farmers' movements have had, and continue to have, in shaping US agrarian politics.[25] Here we will examine some of these efforts towards 'agrarian justice' in the USA and their relationship to food sovereignty. We do not suggest that there exists a consolidated frame of contention, or social movement, for agrarian justice *per se*. Rather, building on Ribot's and Peluso's 'theory of access',[26] we use this term as an umbrella concept to describe the constellation of claims and mobilisations connected to struggles over land, and the ability of producers to benefit from it. By historicising these claims it becomes apparent that the major divides along race and class lines fostered by US farm and land policies have permeated resistance efforts as well. The efforts of USFSA members to bridge some of these divides may be appreciated in this light.

Divisive policies, divided movements

Despite Thomas Paine's calls for 'agrarian justice' as early as 1797,[27] the results of US agrarian policies have been anything but just, from the violent dispossession of Native Americans to stark racial and class disparities in access to land, credit and markets. The Homestead Act of 1862 laid the foundation for such inequalities that persist to this day. The Act proposed to allocate up to 160 acres of farmland to current or prospective US citizens, including women, former slaves and new immigrants.[28] However, slavery would not be abolished until 1865 and, even then, given racial and class inequalities, beneficiaries were mostly white and wealthy.[29] Furthermore, President Andrew Johnson repealed the promise to make 40-acre parcels readily available for former slaves to farm, as had been included in The Freedmen's Bureau Act of 1865.[30] Instead, to sustain the plantation system in the South, former slaves were transitioned into exploitative sharecropping or tenant farming systems.[31]

Despite many barriers, between emancipation and 1910 African Americans came to represent 16.5% of southern agricultural landowners.[32] After reaching a high in 1920 of 925,708 Black farm owners, ongoing discriminatory practices by the US Department of Agriculture (USDA) (see Pigford Class Action Lawsuit[33]), coupled with economic crises, steadily reversed this trend. From 1920 to 2000 the number of Black farmers declined by 98%.[34] In contrast, according to the 2012 US Census of Agriculture, some 96% of all primary farm operators are white and whites hold 98% of privately owned farmland.[35] Under these circumstances resisting Black land loss has been a central point of agrarian struggle that is linked to the broader frame of Black Liberation throughout US history. Agrarian justice for Black farmers has thus meant dismantling the structural racism that permeates US institutions and policies.

The issue of land distribution was also taken up by white farmers at several points in US history, including by predominantly white agrarian populist movements, a lineage that continues to influence agrarian resistance today.[36] Organisations like the Farmers' Alliance at the end of the 19th century called for land reform,[37] and in the 1970s the issue again gained momentum.[38] However, a shift in policy discourse away from 'land reform' towards 'land-use planning', combined with the socialist associations of 'land reform' during the Cold War, contributed to decreased emphasis on land reform by agrarian populists.[39] Meanwhile, during the Great Depression, farms across the country were pushed into crisis. This triggered massive unrest, demanding state support for farmers, resulting in the Agricultural Adjustment Act of 1933. Since then federal policy has played a decisive role in influencing the extent to which farmers are able to receive fair prices.[40] In the face of growing corporate control the issue of fair prices for family farmers has been central to radical agrarian organisations throughout the 20th century and into today.

However, the gradual loss of the redistributive land reform frame among predominantly white agrarian movements may also have helped obscure the inequality embedded in the land tenure system of US agriculture, which provided few opportunities for upward movement between landless and landowning classes.[41] Some have argued that 'agrarian populists missed the opportunity to build alliances across race and class lines by ignoring the struggles of sharecroppers and farmworkers',[42] while others argue that the populists tried to build interracial and interclass coalitions, but ultimately failed.[43] The precarious nature of such alliances among producers can be seen in the makeup and structure of farmers' organisations through history. The Farmers Alliance mentioned above, for example, restricted its membership to whites. When a faction of the Alliance tried to allow Black farmers to join, leadership was divided and finally dealt with the split by allowing a separate but affiliated Colored Farmers' National Alliance and Cooperative Union (CFNACU) to form in 1886.[44]

The collapse of the populist platform paved the way for more class- and race-blind mobilising structures. During the post-New Deal era, in a reflection of farm policy emphasising the production of particular commodity crops, organisations like the National Cattlemen's Association and the National Corn Growers emerged, focusing on maximising commodity production and creating an infrastructure for farmer organising according to crop.[45] McConnell argues that this 'commodityism' de-emphasised class-based organising among farmers.[46]

Convergence: challenges and opportunities

The threats from corporate agribusiness that the populists rallied against continue to this day, as farmers of all races get squeezed out of agriculture. Founded in 1986 in response to the farm crisis, the National Family Farm Coalition (NFFC) has worked to bridge many of the above-mentioned divides. This can be seen in the diversity of its membership, the framing of its work and its approach to alliance building. With grassroots members from 32 states, each with its own local membership base, NFFC estimates that it serves between 20,000 and 25,000 within its membership.[47] From its start NFFC has framed its priorities as including both issues of pricing and corporate control *and* issues of land loss facing farmers of colour. Further, it has sought to articulate the ways in which these issues affect not only farmers, but the broader population. In the mid-1990s NFFC became the first US-based member of La Via Campesina; over time it has adopted food sovereignty as its guiding framework. This has involved articulating among its membership what food sovereignty means to them.[48]

As a founding member of the USFSA, NFFC's work to build a more cohesive movement for food sovereignty in the USA has involved connecting historically embedded struggles in the country – such as struggles against Black land loss and against the corporate consolidation of the food industry – with broader trends, such as the current global rush on farmland. Key to NFFC's repertoire of contention have been policy analysis and proposals. In its proposed Food from Family Farms Act, for instance, citing the history of civil rights struggles, the NFFC advocates 'equitable access to farm and housing programs for all farmers and rural people'.[49] It also addresses what have traditionally been divisive issues, for instance crop subsidies, highlighting the ways in which these are a symptom of broader failed policies that are in fact harming farmers, consumers and food system workers alike. In this way the NFFC offers a voice within the USFSA on the importance of policy reform as part of a strategy for change.

From its founding NFFC's members have included geographically, racially and sectorally diverse farm groups united on the basis of shared (albeit differentiated) class interests.[50] Efforts to further expand its membership continue with the inclusion of fishing and Indigenous groups in recent years. Further, while farmworker groups are not among its members, NFFC has frequently aligned itself with farmworker causes, including actively supporting the campaigns of the Florida-based Coalition of Immokalee Workers (who, in turn, have been advising NFFC members in their respective campaigns). While this broadening has helped diversify perspectives and frames of contention within the organisation, NFFC and other producer groups have been limited in their mobilisation capacity. The population working in the agriculture, forestry, fishing and hunting sector has dwindled to a mere 1.5% of the total workforce today.[51] The types of resistance strategies based on mass mobilisation, which have historically been important drivers of change, are increasingly weakened unless producers continue to build cross-sector alliances. For this reason NFFC's convergence with food justice and immigrant food and farm labour justice organisations through the USFSA represents a strategic opportunity.

Food justice

The food justice movement gained traction, especially in urban areas in the 1990s, drawing inspiration from the racial justice frame of the civil rights and environmental justice movements in the USA, along with concern over the disproportionate impacts of diet-related disease on low-income communities of colour.[52] Many food justice initiatives focus on creating alternative forms of food provision, which include urban agriculture, collective purchasing programmes and community-based markets serving traditionally marginalised areas. The New York-based organisation Just Food describes food justice as 'communities exercising their right to grow, sell, and eat [food that is] fresh, nutritious, affordable, culturally appropriate, and grown locally with care for the well-being of the land, workers and animals'.[53] However, within this frame we find an array of perspectives that highlight the tensions identified by Holt-Giménez and Shattuck,[54] ranging from progressive approaches, where the priority is increasing food access without necessarily dismantling structures causing inequality, to radical approaches that indeed seek to dismantle such structures and see food access as an entry point.

Divisions: depoliticisation vs radicalisation

The divides confronting the food justice movement can best be understood in the context of neoliberal policies that have shaped the mobilising structures and tactics used by US social movements. As neoliberalism has meant a rolling back of the state, the NGO sector has swelled to fill the void. Because food-oriented NGOs span so many different funding categories (eg health, education, human services), data specific to food justice non-profits are unavailable. However, it is clear that the sector continues to grow in the USA. There were roughly 1.58 million non-profits in the USA in 2011, up 21.5% from 2001, and private giving to NGOs passed $300 billion in 2012.[55] A total of 11.4 million people are employed in the non-profit sector in the USA.[56]

As Borras has noted, the role of NGOs in the international food sovereignty movement is a complex and ongoing debate;[57] the same is true regarding US movements. Guthman warns that flows of capital from foundations into food movements ultimately depoliticise the movements, claiming, 'for activist projects, neoliberalisation limits the conceivable because it limits the arguable, the fundable, the organisable'.[58] On one hand, some NGOs serve as important mobilising structures for resistance; on the other, they also need external funds to sustain staff salaries. Thus NGOs of all political orientations tend to frame their work in the desired language of funders. The political agenda of funders and the autonomy given to grantees are therefore key questions in understanding the extent to which channelling resistance through non-profits can depoliticise food movements in the USA. This critique is not new, especially in regard to mobilising for racial justice. Allen describes the motive behind the first Ford Foundation grant to a militant group, the Cleveland chapter of the Congress for Racial Equality (CORE) and its impact on the civil rights movement there in 1967. 'CORE [fit] the bill because its talk about black revolution was believed to appeal to discontented blacks, while its program of achieving black power through massive injections of governmental, business, and Foundation aid seemingly opened the way for continued corporate domination of black communities by means of a new black elite'.[59]

Today the dynamics within one of the largest nationwide food justice networks, the Growing Food and Justice Initiative (GFJI), can be seen as an indication of the divisiveness of the funding climate. The main debate is over the degree to which funders influence and benefit from grantee organisations. A member of the USFSA, GFJI 'is an initiative aimed at dismantling racism and empowering low-income communities of colour through sustainable and local agriculture'.[60] While this mission implies a radical perspective, GFJI's host organisation, Growing Power, has come under fire for accepting a $1 million grant from the Walmart Foundation. As the largest food retailer in the USA, Walmart has been criticised for being 'at the center of the nation's cheap food structure' and for using 'philanthropic donations to push its expansion in urban areas'.[61] However, Growing Power's founder, Will Allen, asserts: 'We can no longer be so idealistic that we hurt the very people we're trying to help'.[62]

For all food justice advocates the problem of food insecurity is a main point of departure, but among the key contributions of the *radical* food justice movement is an analysis of how 'from field to fork, the production and consumption of food is racialized'.[63] Disparities in food access provide a window into a much larger system of race and class relations. Therefore, as Holt-Giménez and Wang argue, 'engaging with the structural aspects of food justice requires addressing race and class in relation to dispossession and control over land, labour, and capital in the food system'.[64]

The history of Black land loss is one starting point for such engagement. During 'The Great Migration', from 1915 to 1960, around five million African Americans migrated from the rural South to urban centres, primarily in the North.[65] This influx of African Americans to the North, claiming voting and other rights of citizenship, ushered in the civil rights movement. Despite some hard-fought gains, however, African American incomes have consistently averaged about 60% of white incomes since World War II.[66] Governance of deeply segregated and marginalised urban Black communities has relied on police brutality, criminalisation and mass incarceration, characterised as 'The New Jim Crow'.[67] Segregation of urban areas has been underwritten by rampant redlining in mortgage lending, with many of the marks of discrimination visible on maps of major cities today.[68] In neighbourhoods like the east and west Oakland flat-lands, incomes tend to be lower and food options largely limited to cheap, highly processed foods.[69] Decreased healthy food access is increasingly linked to the rise in obesity and diet-related diseases in the USA,[70] with the health costs disproportionately born by communities of colour.[71] Furthermore, the intersections of race and class and their connection to food access are not straightforward and merit further research. In St Louis, MO, for instance, pre-dominantly Black communities, regardless of income, are less likely than higher-income white neighbourhoods to have access to healthy foods.[72]

Converging to dismantle racism

Recognising the ways institutional racism is shaping food access, some food jus-tice groups are drawing on the tactics and frames developed during the civil rights movement. Doing so does two things. First, it frames food justice as an extension of the radical anti-racist, anti-capitalist project of Black Liberation, thus rejecting a depoliticised reformist approach. As a member of Soul Fire Farm explains, 'we were attempting to meet a challenge presented to us by Curtis Hayes Muhammad, the veteran civil rights activist: "Recognize that land and food have been used as a weapon to keep black people oppressed [...] Rec-ognize also that land and food are essential to liberation for black people".'[73] In the eyes of youth food justice leader, Anim Steel, 'We can't change the food system by simply changing the tastes and attitudes of regular people any more than the civil rights movement could end segregation without the 1964 Civil Rights Act. Beyond the personal these transformations require political, economic, and cultural changes.'[74]

Second, it opens up food justice activism to alliances with rapidly growing movements against structural racism and state violence such as #BlackLivesMat-ter. These groups in particular are reviving some of the types of disruptive strat-egies of resistance of the civil rights movement, like sit-ins, die-ins, transit stoppages and mass marches. Some groups, like Soul Fire Farm, have already begun bridging the food justice, civil rights and #BlackLivesMatter discourse – an indication that more widespread convergence may further strengthen the radi-cal tendency in the food justice movement. Penniman writes, "state violence is only one among many dangers. The biggest killers of black Americans today are not guns or violence, but diet-related diseases, including heart disease, can-cer, stroke, and diabetes [...] Black youth are well aware that the system does not value their lives.'[75]

The presence of groups such as Soul Fire, Detroit Black Community Food Security Network, GFJI and others in the USFSA is helping to keep anti-racist organising at the forefront of the agenda. Indeed, while explicit mentions of race and racism do not figure very centrally in global food sovereignty discourse, the groups involved in founding the USFSA have found that one cannot talk about food sovereignty in the USA without talking about race. This message was strongly emphasised by people of colour-led organisations in the discussions leading to the founding of the USFSA, sparking a series of challenging and frank dialogues on issues of privilege, leadership and representation that continue to this day. The founding documents of the USFSA reflect this, including that core members must 'commit to take part in racial justice/anti-racism trainings and to actively work to apply anti-racist principles to the work of the Alliance (internally and externally)'.[76] To fulfil these commitments, the USFSA has been following the lead of its food justice members.

Immigrant labour justice

Immigrant workers represent an important and growing part of the US food system. Their struggles intersect with the agrarian justice framework and are also at times discussed as part of the food justice movement.[77] Here, however, we address these mobilisations separately to reflect the fact that historic divisions have kept these workers separate from other food and agrarian movements and the central frame of contention has been one focused on immigrant labour justice.

According to the Food Chain Workers Alliance (FCWA), nearly 20 million 'food chain workers' are currently employed in the USA, as farmworkers, slaughterhouse and other processing facilities workers, warehouse workers, grocery store workers, or restaurant and food service workers.[78] Together these sectors account for over 13% of US GDP.[79] Although specific data are difficult to obtain because of their 'illegal' status, undocumented workers comprise an important part of the labour in the food system and consistently endure lower wages, higher rates of wage theft and food insecurity.[80]

Division by threat of deportation

US labour and immigration policy has served to keep food and especially farm workers isolated and vulnerable in ways that support a low-wage model of food production. First, racial discrimination caused farmworkers to be exempted from basic labour protections in the USA that were instituted from Roosevelt's New Deal onward.[81] Meanwhile, after World War II, programmes like the Mexican Farm Labor Program Agreement of 1942, later called the Bracero Program (until 1964), recruited some two million workers from the Mexican countryside into US fields as temporary guest-workers.[82] US trade policies in the 1990s, namely the North American Free Trade Agreement (NAFTA), and later the Dominican Republic–Central American Free Trade Agreement (CAFTA–DR), led to the displacement of thousands of peasant farmers in Mexico and Central America, significantly increasing new migration flows to the USA.[83] To this day farmworkers in the USA remain excluded from the National Labor Relations Act, which protects the right to organise, and from overtime pay under federal law.[84] Additionally, border policy based on the criminalisation of immigrants has created a flow of workers vulnerable to deportation, for whom the risks of speaking out about workplace violations are even greater than for documented workers. Because of this, many unions in the US labour movement have traditionally perceived immigrant workers as 'unorganisable' and largely excluded them.[85] Some exceptions to this were efforts in the 1930s to organise eastern European immigrants, who were 'welcomed into the house of labor'.[86] In this process eastern Europeans became '"white", shedding their previously racialized status'.[87] This was not so for African Americans, Asians and Latinos, who continued to endure systemic discrimination and economic disadvantage.[88] Such racial disparities persist today. For example, research by FCWA reveals that Black workers are subject to wage theft more than any other racial category and tend to be concentrated in the warehouse sector (76% of the workforce), where the issue is most common. And in grocery Latinos suffer the most wage theft (78.6%).[89]

Beyond the divisions within the labour movement, workers sometimes find themselves overlooked by other food and agrarian mobilisations. When Sinclair published *The Jungle*, revealing the working conditions among Chicago's immigrant workers in meat packing plants at the turn of the 20th century, he complained that the public focused more on the unsanitary conditions of their food than on poor labour standards. He lamented, 'I aimed at the public's heart and by accident hit its stomach'.[90] FCWA has similarly stated that 'the food movement of the last several decades has not focused on sustainable labour practices within the food system, with some notable exceptions...particularly with regard to farmworkers'.[91] Indeed, drawing on the proven tactics of the United Farm Workers of America (UFW), the above-mentioned Coalition of Immokalee Workers has organised tomato pickers in Florida and, allied with consumers, is pressuring corporate buyers to sign 'fair food agreements' aimed at improving both workers' pay and working conditions. These agreements generated $10 million in the first three seasons, to be passed on to workers as increased wages.[92]

Despite their respective links with consumers, farmworkers and small farmers occupy class positions that have historically been in tension. In their survey of 175 organic farmers in California, Shreck et al. found that 47.5% of those who hire labour strongly feel that organic certification should not include criteria about working conditions.[93] Furthermore, 'even if they *believe* that organic agriculture should ensure fair and healthy working conditions for farmworkers, they explain that it is simply not economically viable given the realities of the market'.[94] In response to this deeply entrenched tension, Dolores Huerta of UFW has argued that 'although many small farmers are presently opposed to the farmworker's union, it is in the long run interest of family farmers and farmworkers to join together against big growers and corporations'.[95]

Convergence throughout the food chain

In many ways food sovereignty offers a frame for alliance building between US-based agrarian and immigrant labour movements. At the same time the US case offers some important challenges that the food sovereignty movement may also learn from. Indeed, the question of how farmworkers fit into the vision of food sovereignty remains unresolved,[96] and the role of food workers even less well articulated.

The FCWA is working to address these challenges by actively building bridges between urban and rural workers across the food chain organised via unions, workers' centres and NGOs. While their work is in large part animated by a frame of immigrant labour justice, their participation in the USFSA has helped give food and farm workers a voice in the US food sovereignty movement. The intersection of these movements has helped reveal what a huge (and increasingly mobilised) part of the food system food chain workers represent. Furthermore, the presence of FCWA in the USFSA opens the conversation about what role food and farm workers play in food sovereignty and the unique challenges that immigrant workers deal with, which may present barriers to mobilisation. For instance, Mares et al describe how Latino farmworkers in Vermont have limited mobility for fear of being deported.[97] What does a path towards food sovereignty look like starting from this reality? And how can the movement provide support in these daily struggles? The USFSA has formed an immigration and trade team to begin articulating this aspect of food sovereignty, for instance through its 'Immigration Policy Principles for Food Sovereignty'.[98]

Conclusion

The US food system has historically depended on severely exploited labour. From slaves to share croppers to immigrant food and farm workers, structural racism has served to dehumanise and criminalise those who are most marginalised in the food system. Particularly telling is that today there are more people in the USA behind bars than tending fields.[99] Food sovereignty emerged as a rallying cry for food system transformation from farmers around the world but they cannot do it alone, especially in the USA. Agrarian justice organisations like the NFFC have important proposals and insights and are working tirelessly, but, among other challenges, demographics are not working in their favour. Transformative political impact therefore requires broader mobilisation of more people than there are family farmers. The food justice movement has shown that dismantling racism is a fundamental element of food system change, something that has not been a prominent focus of global food sovereignty organisations for the most part. However, grappling with racism *within* US movements also remains a key challenge, and the continued radicalisation of the food justice movement (ie through integration with #BlackLivesMatter mobilisations) is indeed 'pivotal' in the construction of a transformative platform of convergence for food sovereignty in the USA.[100] The prominence of the NGO sector in the USA further challenges food sovereignty activists and scholars to articulate a vision for the role of NGOs and donor organisations in the movement. Finally, immigrant food and farm workers are actively creating cross-sector bridges and drawing on their numbers to build political power. However, the unique challenges posed by the threat of deportation urge food sovereignty activists to confront the issue of immigration in their visions of the future. The USFSA represents one platform where these struggles are converging and a space for figuring out, through practice, what alliance means.

Admittedly, in trying to explore food sovereignty in the US context, we have perhaps ended up not speaking very much about food sovereignty *per se*. But we suggest that it is delving into unique histories of resistance that may ultimately enable the growth of a more rooted vision of food sovereignty in the USA. Despite the divisions within and between the trajectories of resistance we have highlighted here, lessons for navigating the politics of convergence are also embedded in those fault lines. We would wager that the same is true for movements in other parts of the world, as they shape and are shaped by food sovereignty, and encourage further research in other contexts.

Acknowledgements

Many thanks to Jun Borras, Eric Holt-Giménez, Todd Holmes, Martha Robbins, Salena Tramel and Siena Chrisman for their input and support. And our deepest respect and gratitude to the movements doing the work that has inspired this piece.

Notes

1. "Declaration of Nyéléni."
2. See, for instance, Borras et al., "Transnational Agrarian Movements."
3. Holt-Giménez, *Food Movements Unite!*
4. Desmarais, *La Vía Campesina*.
5. Martínez-Torres and Rosset, "Diálogo de Saberes."
6. Edelman et al., "Introduction," 922.
7. This is not to imply that attempts towards food sovereignty are not complex and fraught with challenges in any setting, including agrarian ones, as described by Bernstein, "Food Sovereignty"; and Agarwal, "Food Sovereignty," among others.
8. Holmes, "Farmer's Market"; Holt-Giménez and Patel, *Food Rebellions!*; Heffernan et al., *Consolidation in the Food and Agriculture System*; and Walker, *The Conquest of Bread*.
9. Pollan, "Voting with your Fork."
10. Trauger, "Toward a Political Geography of Food Sovereignty."
11. Alkon and Agyeman, "Introduction"; Holt Giménez and Shattuck, "Food Crises"; Billings and Cabbil, "Food Justice"; and Figueroa, "Food Sovereignty."
12. Anderson, "The Role of US Consumers."
13. Dickinson, "Beyond the minimally adequate Diet"; and Anguelovski, "Conflicts around Alternative Urban Food Provision."
14. Clendenning and Dressler, "Between Empty Lots and Open Pots."
15. Roman-Alcalá, "Occupy the Farm."
16. Alkon, "Food Justice."
17. Minkoff-Zern, "The New American Farmer"; and Mares et al., "Cultivating Food Sovereignty."
18. Figueroa, "Food Sovereignty," 3.
19. For background on the US Food Sovereignty Alliance and its origins, see Shawki, "The 2008 Food Crisis."
20. Tarrow, *Power in Movement*, 21.
21. Ibid.
22. Garza, "A Herstory of the #BlackLivesMatter Movement."
23. Kamal and Thompson, "Recipe for Decolonization"; Gupta, "Return to Freedom"; and Grey and Patel, "Food Sovereignty as Decolonialization."
24. GRAIN, *Hungry for Land*, 8.
25. Wilson, "Missing Food Movement History."
26. Ribot and Peluso, "A Theory of Access."

27. Paine, *Agrarian Justice*, iii.
28. Geisler, "A History of Land Reform," 13.
29. Ibid., 12; and Barnes, *The People's Land*, xi.
30. Mitchell, *From Reconstruction to Deconstruction*, 20–21.
31. Coates, "The Case for Reparations."
32. Mitchell, *From Reconstruction to Deconstruction*, 21; and Coates, "The Case for Reparations."
33. See https://www.blackfarmercase.com/.
34. Wood and Gilbert, "Returning African American Farmers to the Land," 43.
35. USDA, *2012 Census of Agriculture Race/Ethnicity/Gender Profile*, 13.
36. Wilson, "Missing Food Movement History."
37. Goodwyn, *The Populist Moment*; and Geisler, "A History of Land Reform," 19–20.
38. Barnes, *The People's Land*.
39. Geisler, "A History of Land Reform," 21.
40. Naylor, *Strengthening the Spirit of America*, 8.
41. Kloppenburg and Geisler, "The Agricultural Ladder," 63.
42. McCullen, "Why are all the White Kids sitting Together?," 21.
43. Goodwyn, *The Populist Moment*, 512; and Gerteis, *Class and the Color Line*.
44. Reynolds, *Black Farmers in America*, 5.
45. Mooney and Majka, *Farmers' and Farm Workers Movements*, 97.
46. McConnell, *The Decline of Agrarian Democracy*, 76–81.
47. Griffith, personal communication, 2015.
48. NFFC, *NFFC Food Sovereignty Vision Statement*.
49. NFFC, *Food and Family Farms Act*, 4.
50. See full membership list at http://.net/index.php/who-we-are/nffc-member-groups/.
51. BLS, "Industry Employment and Output Projections to 2022."
52. Patel, "Survival Pending Revolution"; Allen, *Together at the Table*; Herrera et al., "Food Systems and Public Health"; and Bullard, *Unequal Protection*.
53. Cited in Alkon and Agyeman, "Introduction," 5.
54. Holt Giménez and Shattuck, "Food Crises."
55. Pettijohn, *The Nonprofit Sector in Brief*, 1.
56. BLS, "Nonprofits account for 11.4 Million Jobs."
57. Borras, "Reply."
58. Guthman, "Neoliberalism," 1180.
59. Allen, "From Black Awakening," 56.
60. GFJI, "About Us."
61. Fisher and Gottlieb, "Who Benefits?"
62. Ibid.
63. Billings and Cabbil, "Food Justice," 103.
64. Holt-Giménez and Wang, "Reform or Transformation?," 98.
65. Harrison, *Black Exodus*, vii.
66. DeNavas-Walt et al., cited in Patel, "Survival Pending Revolution."
67. Alexander, *The New Jim Crow*.
68. Madrigal, "The Racist Housing Policy."
69. Herrera et al., "Food Systems and Public Health."
70. Levine, "Poverty and Obesity in the US."
71. Treuhaft and Karpyn, *The Grocery Gap*.
72. Baker et al., "The Role of Race and Poverty in Access to Foods," 1.
73. Penniman, "Radical Farmers."
74. Steel, "Youth and Food Justice," 120.
75. Penniman, "Radical Farmers."
76. The founding document of the USFSA is accessible at http://usfoodsovereigntyalliance.org/wp-content/uploads/2013/05/USFSA-Founding-Document-with-date.pdf.
77. Alkon, "Food Justice"; and Gottlieb and Joshi, *Food Justice*.
78. FCWA, *The Hands that Feed Us*.
79. Ibid., 1.
80. Ibid.
81. Linder, "Farm Workers," 1336.
82. Mize and Swords, *Consuming Mexican Labor*, 2.
83. Bacon, *Illegal People*.
84. Smith and Goldberg, *Unity For Dignity*, 16.
85. Milkman, *LA Story*, 9; and Zolberg, "Rethinking the Last 200 Years."
86. Milkman, *Organizing Immigrants*, 6.
87. Ibid., 5.
88. Ibid., 6.
89. FCWA, *The Hands that Feed Us*, 26.

90. Ibid., 13.
91. Ibid., 16.
92. CIW, "About CIW, Coalition of Immokalee Workers."
93. Shreck et al., "Social Sustainability," 444. Emphasis in original.
94. Ibid.
95. Paraphrased by Barnes, *The People's Land*, 218.
96. Fairbairn, "Framing Transformation," 226; and Patel, "Food Sovereignty," 667.
97. Mares et al., "Cultivating Food Sovereignty," 10.
98. USFSA, *Immigration Policy Principles*.
99. NAACP, "Criminal Justice Fact Sheet"; and USDA, *2012 Census of Agriculture United States Summary and State Data*.
100. Holt-Giménez and Wang, "Reform or Transformation?"

Bibliography

Agarwal, Bina. "Food Sovereignty, Food Security and Democratic Choice: Critical Contradictions, Difficult Conciliations." *Journal of Peasant Studies* 41, no. 6 (2014): 1247–1268. doi:10.1080/03066150.2013.876996.

Alexander, Michelle. *The New Jim Crow: Mass Incarceration in the Age of Colorblindness*. New York: New Press, 2010.

Alkon, Alison Hope. "Food Justice, Food Sovereignty and the Challenge of Neoliberalism." Conference Paper #38. Yale University, 2013. http://www.yale.edu/agrarianstudies/foodsovereignty/pprs/38_Alkon_2013.pdf.

Alkon, Alison Hope, and Julian Agyeman. "Introduction: The Food Movement as Polyculture." In *Cultivating Food Justice: Race, Class, and Sustainability*, edited by Alison Hope Alkon and Julian Agyeman, 1–20. Cambridge, MA: MIT Press, 2011.

Allen, Patricia. *Together at the Table: Sustainability and Sustenance in the American Agrifood System*. University Park, PA: Penn State University Press, 2004.

Allen, Robert L. "From Black Awakening in Capitalist America." In *The Revolution will not be Funded*, edited by INCITE Women of Color Against Violence, 53–63. Cambridge, MA: South End Press, 2007.

Anderson, Molly D. "The Role of US Consumers and Producers in Food Sovereignty." Conference Paper #31. Yale University, 2013. http://www.yale.edu/agrarianstudies/foodsovereignty/pprs/31_Anderson_2013.pdf.

Anguelovski, Isabelle. "Conflicts around Alternative Urban Food Provision: Contesting Food Privilege, Food Injustice, and Colorblindness in Jamaica Plain, Boston." Conference Paper #84. Institute of Social Studies (ISS), The Hague, 2014. http://www.iss.nl/fileadmin/ASSETS/iss/Research_and_projects/Research_net works/ICAS/84_Anguelovski.pdf

Bacon, David. *Illegal People: How Globalization Creates Migration and Criminalizes Immigrants*. Boston, MA: Beacon Press, 2008.

Baker, Elizabeth A., Mario Schootman, Ellen Barnidge, and Cheryl Kelly. "The Role of Race and Poverty in Access to Foods That Enable Individuals to Adhere to Dietary Guidelines." *Preventing Chronic Disease* 3, no. 3 (2006): A76.

Barnes, Peter, ed. *The People's Land: A Reader on Land Reform in the United States*. Emmaus, PA: Rodale Press/National Coalition for Land Reform, 1975.

Bernstein, Henry. "Food Sovereignty via the 'Peasant Way': A Sceptical View." *Journal of Peasant Studies* 41, no. 6 (2014): 1031–1063. doi:10.1080/03066150.2013.852082.

Billings, David, and Lila Cabbil. "Food Justice: What's Race got to do with It?" *Race/Ethnicity: Multidisci-plinary Global Contexts* 5, no. 1 (2011): 103–112. doi:10.2979/racethmulglocon.5.1.103.

BLS. "Industry Employment and Output Projections to 2022." *Monthly Labor Review*. US Bureau of Labor Statistics, December 2013. http://www.bls.gov/opub/mlr/2013/article/industry-employment-and-output-pro jections-to-2022-1.htm.

BLS. "Nonprofits account for 11.4 Million Jobs, 10.3 percent of All Private Sector Employment." *The Eco-nomics Daily*. US Bureau of Labor Statistics, October 21, 2014. http://www.bls.gov/opub/ted/2014/ted_20141021.htm.

Borras, Saturnino M. "Reply: Solidarity – Re-examining the 'Agrarian Movement–NGO' Solidarity Relations Discourse." *Dialectical Anthropology* 32, no. 3 (2008): 203–209. doi:10.1007/s10624-008-9068-3.

Borras, Saturnino M., Marc Edelman, and Cristóbal Kay. "Transnational Agrarian Movements: Origins and Politics, Campaigns and Impact." In *Transnational Agrarian Movements Confronting Globalization*, edited by Saturnino M Borras, Marc Edelman and Cristóbal Kay, 1–36. Chichester: Wiley-Blackwell, 2008.

Bullard, Robert D., and Unequal Protection. *Environmental Justice & Communities of Color*. San Francisco: Random House, 1994.

CIW. "About CIW, Coalition of Immokalee Workers." 2015. http://ciw-online.org/about/.

Clendenning, Jessica, and Wolfram Dressler. "Between Empty Lots and Open Pots –Understanding the Rise of Urban Food Movements in the USA." Conference Paper #48. Yale University, 2013.

Coates, Ta-Nehisi. "The Case for Reparations." *The Atlantic*, June 2014. http://www.theatlantic.com/galleries/reparations/1/.

"Declaration of Nyéléni," 2007. http://www.nyeleni.org/spip.php?article290.

Desmarais, Annette Aurélie. *La Vía Campesina: Globalization and the Power of Peasants*. Halifax/London: Fernwood Publishing/Pluto Press, 2007.

Dickinson, Maggie. "Beyond the minimally adequate Diet: Food Stamps and Food Sovereignty in the US." Conference Paper #23. Program in Agrarian Studies, Yale University, 2013. http://www.yale.edu/agrarian studies/foodsovereignty/pprs/23_Dickinson_2013.pdf

Edelman, Marc, Tony Weis, Amita Baviskar, Saturnino M. Borras, Eric Holt-Giménez, Deniz Kandiyoti, and Wendy Wolford. "Introduction: Critical Perspectives on Food Sovereignty." *Journal of Peasant Studies* 41, no. 6 (2014): 911–931. doi:10.1080/03066150.2014.963568.

Fairbairn, Madeleine. "Framing Transformation: The Counter-hegemonic Potential of Food Sovereignty in the US Context." *Agriculture and Human Values* 29, no. 2 (2012): 217–230. doi:10.1007/s10460-011-9334-x.

FCWA. *The Hands that Feed Us: Challenges and Opportunities for Workers along the Food Chain*. Los Angeles, CA: Food Chain Workers Alliance, June 6, 2012.

Figueroa, Meleiza. "Food Sovereignty in Everyday Life: Toward a People-centered Approach to Food Systems." *Globalizations* (2015): 1–15. doi:10.1080/14747731.2015.1005966.

Fisher, Andy, and Robert Gottlieb. "Who Benefits when Walmart Funds the Food Movement?" *Civil Eats*, December 18, 2014. http://civileats.com/2014/12/18/who-benefits-when-walmart-funds-the-food-movement/.

Garza, Alicia. "A Herstory of the #BlackLivesMatter Movement by Alicia Garza" *Feminist Wire*, Accessed February 8, 2015. http://thefeministwire.com/2014/10/blacklivesmatter-2/

Geisler, Charles. "A History of Land Reform in the United States." In *Land Reform, American Style*, edited by Charles Geisler and Frank Popper, 7–34. Totowa, NJ: Rowman & Allanheld, 1984.

Gerteis, Joseph. *Class and the Color Line: Interracial Class Coalition in the Knights of Labor and the Populist Movement*. Durham, NC: Duke University Press, 2007.

GFJI. "About Us." Growingfoodandjustice. Accessed February 2, 2015. http://www.growingfoodandjustice.org/About_Us.html

Goodwyn, Lawrence. *The Populist Moment*. New York, NY: Oxford University Press, 1978.

Gottlieb, Robert, and Anupama Joshi. *Food Justice*. Cambridge, MA: MIT Press, 2010.

GRAIN. *Hungry for Land*, May 2014. http://www.grain.org/article/entries/4929-hungry-for-land-small-farmers-feed-the-world-with-less-than-a-quarter-of-all-farmland

Grey, Sam, and Raj Patel. "Food Sovereignty as Decolonialization: Some Contributions from Indigenous Movements to Food System and Development Politics." *Agriculture and Human Values* (2014): doi:10.1007/s10460-014-9548-9.

Gupta, Claire. "Return to Freedom: Anti-GMO Aloha 'Aina Activism on Molokai as an Expression of Place-based Food Sovereignty." *Globalizations*. September (2014). Forthcoming. doi: 10.1080/14747731.2014.957586

Guthman, Julie. "Neoliberalism and the Making of Food Politics in California." *Geoforum* 39, no. 3 (2008): 1171–1183. doi:10.1016/j.geoforum.2006.09.002.

Harrison, Alferdteen. *Black Exodus: The Great Migration from the American South*. Jackson, MS: University Press of Mississippi, 1991.

Heffernan, William, Hendrickson Mary, and Gronski, Robert. *Consolidation in the Food and Agriculture System*. Report to the National Farmers Union, February 5, 1999. http://www.foodcircles.missouri.edu/whstudy.pdf.

Herrera, Henry, Navina, Khanna, and Davis, Leon. "Food Systems and Public Health: The Community Perspective." *Journal of Hunger & Environmental Nutrition* 4, nos. 3–4 (2009): 430–445. doi:10.1080/19320240903347446.

Holmes, Todd. "Farmer's Market: Agribusiness and the Agrarian Imaginary in the California Far West." *California History* 90, no. 2 (2013): 24–74.

Holt-Giménez, Eric, ed. *Food Movements Unite! Strategies to Transform our Food Systems*. Oakland, CA: Food First Books, 2011.

Holt-Giménez, Eric, and Raj Patel. *Food Rebellions! Crisis and the Hunger for Justice*. Oakland, CA: Food First Books/Fahumu Books/Grassroots International, 2009.

Holt Giménez, Eric, and Annie Shattuck. "Food Crises, Food Regimes and Food Movements: Rumblings of Reform or Tides of Transformation?" *Journal of Peasant Studies* 38, no. 1 (2011): 109–144. doi:10.1080/03066150.2010.538578.

Holt-Giménez, Eric, and Yi Wang. "Reform or Transformation? The Pivotal Role of Food Justice in the US Food Movement." *Race/Ethnicity* 5, no. 1 (2011): 83–102.

Kamal, Asfia Gulrukh, and Thompson, Shirley. Recipe for Decolonization and Resurgence: Story of O-Pipon-Na-Piwin Cree Nation's Indigenous Food Sovereignty Movement. Conference Paper #46, Yale University, 2013. http://www.yale.edu/agrarianstudies/foodsovereignty/pprs/46_Kamal_Thompson_2013.pdf.

Kloppenburg, Jack, and Charles Geisler. "The Agricultural Ladder: Agrarian Ideology and the Changing Structure of US Agriculture." *Journal of Rural Studies* 1, no. 1 (1985): 59–72.

Levine, J. A. "Poverty and Obesity in the US." *Diabetes* 60, no. 11 (2011): 2667–2668. doi:10.2337/db11-1118.

Linder, Marc. "Farm Workers and the Fair Labor Standards Act: Racial Discrimination in the New Deal." *Texas Law Review* 65 (1986): 1335–1395.

Madrigal, Alexis C. "The Racist Housing Policy that made your Neighborhood." *The Atlantic*, May 22, 2014. http://www.theatlantic.com/business/archive/2014/05/the-racist-housing-policy-that-made-your-neighborhood/371439/.

Mares, Teresa M., Naomi Wolcott-MacCausland, and Jessie Mazar. "Cultivating Food Sovereignty where there are few Choices." Conference Paper #29. Yale University, 2013. http://www.yale.edu/agrarianstudies/food sovereignty/pprs/29_Mares_2013.pdf.

Martínez-Torres, María Elena, and Peter M. Rosset. "Diálogo de Saberes in La Vía Campesina: Food Sovereignty and Agroecology." *Journal of Peasant Studies* 41, no. 6 (2014): 979–997. doi:10.1080/03066150.2013.872632.

McConnell, Grant. *The Decline of Agrarian Democracy*. New York: Atheneum, 1969.

McCullen, Christie Grace. "Why are all the White Kids sitting together in the Farmers Market? Whiteness in the Davis Farmers Market and Alternative Agrifood Movement." MSc diss., University of California, Davis, 2008.

Milkman, Ruth. *LA Story: Immigrant Workers and the Future of the US Labor Movement*. New York: Russell Sage Foundation, 2006.

Milkman, Ruth. *Organizing Immigrants: The Challenge for Unions in Contemporary California*. Ithaca, NY: Cornell University Press, 2000.

Minkoff-Zern, Laura-Anne. "The New American Farmer: The Agrarian Question, Food Sovereignty and Immigrant Mexican Growers in the United States." Conference Paper #16. Yale University, 2013. http://bio www2.biology.yale.edu/agrarianstudies/foodsovereignty/pprs/16_MinkoffZern_2013.pdf.

Mitchell, Thomas W. *From Reconstruction to Deconstruction: Undermining Black Landownership, Political Independence and Community through Partition Sales of Tenancies in Common*. Madison, WI: Land Tenure Center, University of Wisconsin-Madison, 2000.

Mize, Ronald L., and Alicia C.S. Swords. *Consuming Mexican Labor: From the Bracero Program to NAFTA*. Toronto: University of Toronto Press, 2010.

Mooney, Patrick H., and Theo J. Majka. *Farmers' and Farm Workers Movements: Social Protest in American Agriculture*. Social Movements Past and Present. New York, NY: Twayne Publishers, 1995.

NAACP. "Criminal Justice Fact Sheet." National Association for the Advancement of Colored People, 2015. http://www.naacp.org/pages/criminal-justice-fact-sheet.

Naylor, George. *Strengthening the Spirit of America*. St. Louis, MO: Farm Aid, 1986.

NFFC. *Food and Family Farms Act: A Proposal for the 2007 US Farm Bill*. Washington, DC: National Family Farm Coalition, 2006.

NFFC. *NFFC Food Sovereignty Vision Statement*. National Family Farm Coalition, n.d. http://nffc.net/Farmers %20Worldwide/Food%20Sovereignty%20One-Pager.pdf.

Paine, Thomas. *Agrarian Justice*. Alex Catalogue, 1999. http://schalkenbach.org/library/henry-george/grun dskyld/pdf/p_agrarian-justice.pdf.

Patel, Raj. "Food Sovereignty." *Journal of Peasant Studies* 36, no. 3 (2009): 663–706. doi:10.1080/03066150903143079.

Patel, Raj. "Survival Pending Revolution: What the Black Panthers can teach the US Food Movement." In *Food Movements Unite! Strategies to Transform Our Food Systems*, edited by Eric Holt-Giménez, 115–136. Oakland, CA: Food First Books, 2011.

Penniman, Leah. "Radical Farmers use Fresh Food to fight Racial Injustice and the New Jim Crow." *YES! Magazine*, 2015. http://yesmagazine.org/peace-justice/radical-farmers-use-fresh-food-fight-racial-injustice-black-lives-matter.

Pettijohn, Sarah L. *The Nonprofit Sector in Brief: Public Charities, Giving, and Volunteering, 2013*. Washington, DC: Urban Institute, 2013.

Pollan, Michael. "Voting with your Fork." *New York Times*, online edition, May 7, 2006. http://pollan.blogs.ny times.com/2006/05/07/voting-with-your-fork/?_r=0.

Reynolds, Bruce J. *Black Farmers in America, 1865–2000: The Pursuit of Independent Farming and the Role of Cooperatives*. RBS Research. Rural Business–Cooperative Service, US Department of Agriculture, October 2002.

Ribot, Jesse C., and Nancy Lee Peluso. "A Theory of Access." *Rural Sociology* 68, no. 2 (2003): 153–181.

Roman-Alcalá, Antonio. "Occupy the Farm: A Study of Civil Society Tactics to Cultivate Commons and Construct Food Sovereignty in the United States." Conference Paper #75. Yale University, 2013. http://www.yale.edu/agrarianstudies/foodsovereignty/pprs/75_Roman_Alcala_2013.pdf.

Shawki, Noha. "The 2008 Food Crisis as a Critical Event for the Food Sovereignty and Food Justice Movements." APSA 2011 Annual Meeting Paper, 2011. http://ijsaf.org/archive/19/3/shawki.pdf.

Shreck, Aimee, Christy Getz, and Gail Feenstra. "Social Sustainability, Farm Labor, and Organic Agriculture: Findings from an Exploratory Analysis." *Agriculture and Human Values* 23, no. 4 (2006): 439–449. doi:10.1007/s10460-006-9016-2.

Smith, Rebecca, and Harmony Goldberg. "Unity for Dignity: Expanding the Right to Organize to win Human Rights at Work." Paper presented at the Excluded Workers Congress, December 2010. http://www.united workerscongress.org/uploads/2/4/6/6/24662736/ewc_rpt_final4.pdf.

Steel, Anim. "Youth and Food Justice: Lessons from the Civil Rights Movement." In *Food Movements Unite! Strategies to Transform our Food Systems*, edited by Eric Holt-Giménez, 120. Oakland: Food First Books, 2011.

Tarrow, S. G. *Power in Movement: Social Movements and Contentious Politics*. 2nd ed. Cambridge: Cambridge University Press, 1998.

Trauger, Amy. "Toward a Political Geography of Food Sovereignty: Transforming Territory, Exchange and Power in the Liberal Sovereign State." *Journal of Peasant Studies* 41, no. 6 (2014): 1131–1152. doi:10.1080/03066150.2014.937339.

Treuhaft, Sarah, and Karpyn, Allison . *The Grocery Gap: Who has Access to Healthy Food and Why it Matters*. Oakland, CA: Policy Link/The Food Trust, 2009.

USDA. *2012 Census of Agriculture Race/Ethnicity/Gender Profile*. US Department of Agriculture, National Agricultural Statistics Service, 2012. http://agcensus.usda.gov/Publications/2007/Online_Highlights/County_Profiles/Vermont/cp50007.pdf.

USDA. *2012 Census of Agriculture: United States Summary and State Data*. Geographic Area Series. United States Department of Agriculture, 2014.

USFSA. *Immigration Policy Principles for Food Sovereignty*, 2013. http://usfoodsovereigntyalliance.org/immigration-policy-principles-for-food-sovereignty/.

Walker, Richard A. *The Conquest of Bread: 150 Years of Agribusiness in California*. New York: The New Press, 2004.

Wilson, Brad. "Missing Food Movement History: Highlights of Family Farm Justice, 1950–2000." *La Vida Locavore*, February 26, 2012. http://www.lavidalocavore.org/diary/5106/missing-food-movement-history-highlights-of-family-farm-justice-19502000.

Wood, Spencer D., and Jess Gilbert. "Returning African American Farmers to the Land – Recent Trends and a Policy Rationale." *Review of Black Political Economy* 27, no. 4 (2000): 43–64.

Zolberg, Aristide. "Rethinking the Last 200 Years of US Immigration Policy." *Migration Information Source*, Feature, June 1, 2006. http://www.migrationpolicy.org/article/rethinking-last-200-years-us-immigration-policy.

Index

INDEX

INDEX

Milton Keynes UK
Ingram Content Group UK Ltd.
UKHW051854071024
449327UK00025B/1947

9 780367 110383